HOW TO TEACH MATHEMATICS

a personal perspective

Steven G. Krantz

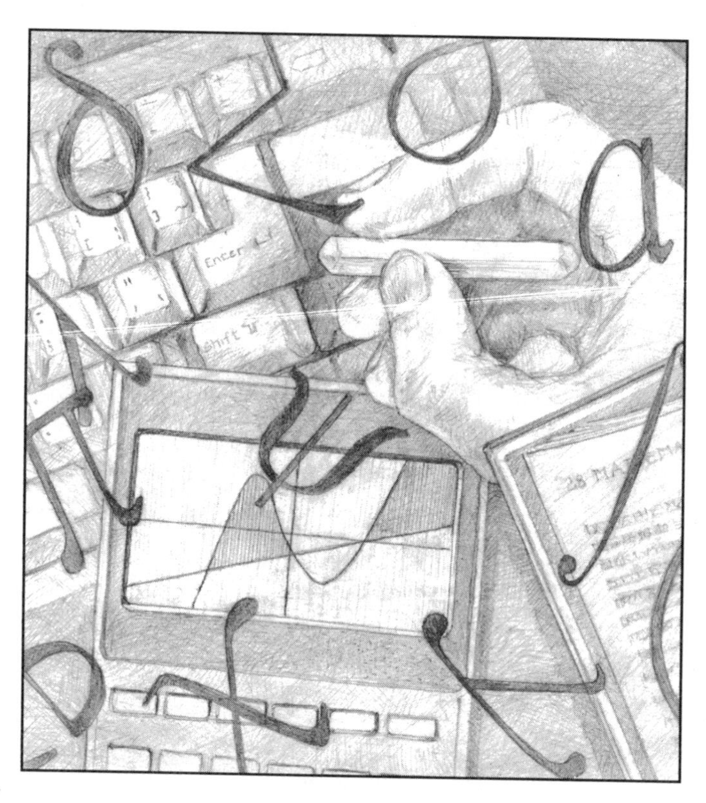

American Mathematical Society
Providence, Rhode Island

1991 *Mathematics Subject Classification*. Primary 00A25, 00A05, 00A20.

Library of Congress Cataloging-in-Publication Data

Krantz, Steven G. (Steven George), 1951–
How to teach mathematics: a personal perspective/Steven G. Krantz.
p. cm.
Includes bibliographical references and index.
ISBN 0-8218-0197-X
1. Mathematics—Study and teaching. I. Title.
QA11.K776 1994 93-19653
510′.71′1–dc20 CIP

Printed in the United States of America

The paper used in this book is acid-free and falls within the guidelines
established to ensure permanence and durability. ♾

♻ Printed on recycled paper.

This publication was prepared by the author using $\mathcal{A}\mathcal{M}\mathcal{S}$-TEX,
the American Mathematical Society's TEX macro system.

10 9 8 7 6 5 4 3 2   98 97 96 95 94

To Robert L. Borrelli, teacher and friend.

Table of Contents

Preface

While most mathematics instructors prepare their lectures with care, and endeavor to do a creditable job at teaching, their ultimate effectiveness is shaped by their attitudes. As an instructor ages (and I speak here of myself as much as anyone), he finds that he is less in touch with his students, that a certain ennui has set in, and (alas) perhaps that teaching does not hold the allure and sparkle that it once had. Depending on the sort of department in which he works, he may also feel that hotshot researchers and book writers get all the perks and that "mere teachers" are viewed as drones.

As a result of this fatigue of enthusiasm, a professor will sometimes prepare for a lecture *not* by writing some notes or by browsing through the book but by lounging in the coffee room with his colleagues and bemoaning (a) the shortcomings of the students, (b) the shortcomings of the text, and (c) that professors are overqualified to teach calculus. Fortified by this yoga, the professor will then proceed to his class and give a lecture ranging from dreary to arrogant to boring to calamitous. The self-fulfilling prophecy having been fulfilled, the professor will finally join his cronies for lunch and be debriefed as to (a) the shortcomings of the students, (b) the shortcomings of the text, and (c) that professors are overqualified to teach calculus.

There is nothing new in this. The aging process seems to include a growing feeling that the world is going to hell on a Harley. A college teacher is in continual contact with young people; if he/she feels ineffectual or alienated as a teacher, then the unhappiness can snowball.

Unfortunately, the sort of tired, disillusioned instructors that I have just described exist in virtually every mathematics department. A college teacher who just doesn't care anymore is a poor role model for the novice instructor. Yet that novice must turn somewhere to learn how to teach. You cannot learn to play the piano or to ski by watching someone else do it. And the fact of having sat in a classroom for most of your life does not mean that you know how to teach.

The purpose of this booklet is to set down the traditional principles of good teaching in mathematics—as viewed by this author. While perhaps most experienced mathematics instructors would agree with much of what is in this booklet, in the final analysis this tract must be viewed as a personal polemic on how to teach.

Teaching is important. University administrations, from the top down, are today holding professors accountable for their teaching. Both in tenure and promotion decisions and in the hiring of new faculty, mathematics (and other) departments must make a case that the candidate is a capable and talented teacher. In some departments at Harvard, a job candidate must now present a "teaching dossier" as well as an academic dossier. It actually happens that good mathematicians who are really rotten teachers do not get that promotion or do not get tenure or do not get the job that they seek.

The good news is that it requires no more effort, no more preparation, and no more time to be a good teacher than to be a bad teacher. The proof is in this booklet. Put in other words, this booklet is not written by a true believer who is going to exhort you to dedicate every waking hour to learning your students' names and designing seating charts. On the contrary, this booklet is written by a pragmatist who values his time and his professional reputation, but is also considered to be rather a good teacher.

I intend this booklet primarily for the graduate student or novice instructor preparing to sally forth into the teaching world; but it also may be of some interest to those who have been teaching for a few or even for several years. As with any endeavor that is worth doing well, teaching is one that will improve if it is subjected to periodic re-examination.

Let me begin by drawing a simple analogy: By the time you are a functioning adult in society, the basic rules of etiquette are second nature to you. You know instinctively that to slam a door in someone's face is (i) rude, (ii) liable to invoke reprisals, and (iii) not likely to lead to the making of friends and the influencing of people. The keys to good teaching are at approximately the same level of obviousness and simplicity. But here is where the parallel stops. We are all *taught* (by our parents) the rules of behavior when we are children. Traditionally, we (as mathematicians) are not taught anything, when we are undergraduate or graduate students, about what comprises effective teaching.

In the past we have assumed that either

(i) Teaching is unimportant.

or

(ii) The components of good teaching are obvious.

or

(iii) The budding professor has spent a lifetime sitting in front of professors and observing teaching, both good and bad; surely, therefore, this person has made inferences about what traits define an effective teacher.

I have already made a case that (i) is false. I agree wholeheartedly with (ii). The rub is (iii). If proof is required that at at least some mathematicians have given little thought to exposition and to teaching, then think of the last several colloquia that you have heard. How many were good? How many were inspiring? This is supposed to be the stuff that matters—getting up in front of our peers and touting our theorems. Why is it that people who have been doing it for twenty or thirty years still cannot get it right? Again, the crux is item (iii) above. There are some things that we do not learn by osmosis. How to lecture and how to teach are among these.

Of course the issue that I am describing is not black and white. If there were tremendous peer support in graduate school and in the professorial ranks for great teaching, then we would force ourselves to figure out how to teach well. But often there is not. The way to make points in graduate school is to ace the qualifying exams and then to write an excellent thesis. It is unlikely that your thesis advisor wants to spend a lot of time with you chatting about how to teach the chain rule. After all, he/she has tenure and is probably more worried about where his/her next theorem or next grant or next raise is coming from than about such prosaic matters as calculus.

The purpose of this booklet is to prove that good teaching requires relatively little effort (when compared with the alternative), will make the teaching process a positive part of your life, and can earn you the respect of your colleagues. In large part I will be stating the obvious to people who, in theory, already know what I am about to say.

It is possible to argue that we are all wonderful teachers, simply by *fiat*, but that the students are too dumb to appreciate us. Saying this, or thinking it, is analogous to proposing to reduce crime in the streets by widening the sidewalks. It is double-talk. If you are not transmitting knowledge, then you are not teaching. We are not hired to train the ideal platonic student. We are hired to train the particular students who attend our particular universities. It is our duty to learn how to do so.

This is a rather personal document. After all, teaching is a rather personal activity. But I am not going to advise you to tell jokes in your classes, or to tell anecdotes about mathematicians, or to dress like Gottfried Wilhelm von Leibniz when you teach the product rule. Many of these techniques only work for certain individuals, and only in a form suited to those individuals. Instead I wish to distill out, in this booklet, some universal truths about the teaching of mathematics. I also want to go beyond the platitudes that you will find in books about teaching *all* subjects (such as "type all your exams", "grade on a bell-shaped curve") and talk about issues that arise specifically in the teaching of mathematics. I want to talk about principles of teaching that will be valid for all of us.

My examples are drawn from the teaching of courses ranging from calculus to real analysis and beyond. Lower division courses seem to be an ideal crucible in which to forge teaching skills, and I will spend most of my time commenting on those. Upper division courses offer problems of their own, and I will say a few words about those. Graduate courses are dessert. You figure out how you want to teach your graduate courses.

There are certainly differences, and different issues, involved in teaching every different course; the points to be made in this booklet will tend to transcend the seams and variations among different courses. If you do not agree in every detail with what I say, then I hope that at least my remarks will give you pause for thought. In the end, you must decide for yourself what will take place in your classroom.

There is a great deal of discussion these days about developing new ways to teach mathematics. I'm all for it. So is our government, which is generously funding many "teaching reform" projects. However, the jury is still out regarding which of these new methods will prove to be of lasting value. It is not clear yet exactly how MATHEMATICA notebooks or computer algebra systems or interactive computer simulations should be used in the lower division mathematics classroom. Given that a large number of students need to master a substantial amount of calculus during the freshman year, and given the limitations on our resources, I wonder whether alternatives to the traditional lecture system—such as, for instance, Socratic dialogue—are the correct method for getting the material across. Every good new teaching idea should be tried. Perhaps in twenty years some really valuable new techniques will have evolved. They do not seem to have evolved yet.

In 1993 I must write about methods that I know and that I have found to be effective. Bear this in mind: experimental classes are experimental. They usually lie outside the regular curriculum. It will be years before we know for sure whether students taught with the new techniques are understanding and retaining the material satisfactorily and are going on to successfully complete their training. Were I to write about some of the experimentation currently being performed then this book would of necessity be tentative and inconclusive.

There are those who will criticize this book for being reactionary. I welcome their remarks. I have taught successfully, using these methods, for twenty years. Using critical self examination, I find that my teaching gets better and better, my students appreciate it more, and (most importantly) it is more and more effective. I cannot in good conscience write of unproven methods that are still being developed and that have not stood the test of time. I leave that task for the advocates of those methods.

In fact I intend this book to be rather prescriptive. The techniques that I discuss here are ones that have been used for a long time. They work. Picasso's revolutionary techniques in painting were based on a solid classical foundation. By analogy, I think that before you consider new teaching techniques you should acquaint yourself with the traditional ones. Spending an hour or two with this booklet will enable you to do so.

I am grateful to the Fund for the Improvement of Post-Secondary Education for support during a part of the writing of this book. Randi D. Ruden read much of the manuscipt critically and made decisive contributions to the clarity and precision of many passages. Josephine S. Krantz served as a valuable assistant in this process. Bruce Reznick generously allowed me to borrow some of the ideas from his booklet *Chalking It Up*. I also thank Dick Askey, Brian Blank, Bettye Anne Case, Joe Cima, John Ewing, Mark Feldman, Jerry Folland, Ron Freiwald, Paul Halmos, Gary Jensen, John McCarthy, Alec Norton, Mark Pinsky, Bruce Reznick, Richard Rochberg, Bill Thurston, and the students in our teaching seminar at Washington University for many incisive remarks on different versions of the manuscript. The publications committees of the Mathematics Association of America and of the American Mathematical Society have provided me with detailed reviews and valuable advice for the preparation of the final version of this book.

SGK

CHAPTER I

Guiding Principles

1. Respect

You cannot be a good teacher if you do not respect yourself. If you are going to stand up in front of thirty people or three hundred people and try to teach them something, then you had better

- Believe that you are well qualified to do so.
- Want to do so.
- Be *prepared* to do so.
- Make sure that these characteristics are evident to your audience.

It is a privilege to stand before a group of people—whether they be young adults or your own peers—and to share your thoughts with them. You should acknowledge this privilege by (a) dressing appropriately for the occasion, (b) making an effort to communicate with your audience, (c) respecting the audience's point of view.

One completely obvious fact is this: you should have your material *absolutely mastered* before you enter the classroom. If you *do* possess this mastery, then you can expend the majority of your effort and attention on conveying the material to the audience. If, instead, you have a proof or an example that is not quite right, and if you stand in front of the group trying to fix it, then you will lose all but the die-hards quickly.

It is easy to rationalize that if the students were more able then they could roll with the ups and downs of your lecture. This is strictly illogical. How do you behave when you are listening to a colloquium or seminar and the lecturer goes off into orbit—either to fix an incorrect argument or into a private conversation with his buddy in the front row or, worse, into a private conversation with himself? All right then, now that you have admitted honestly how *you* behave, then can you really expect unseasoned freshmen to be tolerant when you do not seem to be able to do the examples that *they* are expected to do?

One of the best arguments for even elementary college mathematics courses to be taught by people with advanced degrees is this: since the material is all trivial and obvious to the august professor then the professor can maintain a broad sense of perspective, will not be thrown by questions, and can concentrate on the act of *teaching.*

If you respect yourself then, it follows logically, you will respect your audience: You should prepare your lecture. That way you will not be surprised by gaps in your thinking, you will not have to cast around for a necessary idea, you will not lose your train of thought in the lecture.

Throughout this booklet, I will repeatedly exhort you to prepare your lectures. I do not necessarily have in mind that you should spend an inordinate amount of *time* preparing. Consider by analogy the psychology of sport. Weight lifters, for example, are taught to meditate in a certain fashion before a big lift. Likewise, preparing is a way to collect your thoughts and put yourself in the proper frame of mind to give a lecture. It makes good sense. See also Section 1.4.

To me, preparation is the core of effective teaching. While this may sound like a tautology, and not worth developing, there is in fact more here than meets the eye. Just as being a bit organized relieves you of the stress and nuisance of spending hours looking for a postage stamp or a pair of scissors when you need them, so being the master of your subject gives you the ability to cope with the unexpected, to handle questions creatively, and to give proper stimulus to your students. An experienced and knowledgeable teacher who is comfortable with his/her craft is constantly adjusting the lecture, in real time, to suit the expressions on the students' faces, to suit their responses to queries and prods, to suit the rate and thoroughness with which they are absorbing the material. Just as a good driver is constantly making little adjustments in steering in response to road conditions, weather, and so forth, it is also the case that a good teacher (just as unconsciously) is engaging in a delicate give and take with the audience. Complete mastery is the unique tool that gives you the freedom to develop this skill.

You should treat questions with respect. I go into every class that I teach knowing full well that I am probably much smarter, and certainly much better informed, than most of the people in the room. But I do not need to use a room full of eighteen year olds as a vehicle for bolstering my ego. If a student asks a question, even a stupid one, then I treat it as an event. A wrong question can be turned into a good one with a simple turn of phrase from the instructor. If the question requires a lengthy answer then give a short one and encourage the student to see you after class. If you insult—even gently—the questioner then you not only offend that person but perhaps everyone else in the room. Once the students have turned hostile it is difficult to win them back, both on that day and on subsequent days.

If a student asks for permission to hand a paper in late, and you are tempted to say "I've set the deadline and the deadline is the deadline and how dare you ask me for an extension," then you should pause. You should ask yourself what you will have accomplished with this little speech. Does it really make any difference if this particular student hands the homework paper in tomorrow morning? If it does not, then is there any reason not to grant the extension? Is it possible that, given a little more time, this student will learn something extra? This consideration is just showing the student the sort of respect that you would have liked to have been shown when you were a student.

If a student wants to discuss why an exam was graded the way that it was, or why so many points were deducted on problem 5, or why he/she is not doing as well in the class as anticipated, then don't hide behind your rank. If you are not prepared to say a sentence or two about why you graded problem 5 the way that you did, it probably means that you graded the problem very sloppily and are afraid that you will be made to look foolish. It may mean that you are uncomfortable with confrontational scenes. But confrontational scenes may be avoided almost 100% of the time. If you are looking over problem 5 with the student, and if it turns out that the problem was basically right but you only gave 3 points out of 10, then you might say "I guess I read this problem too quickly. I get bleary-eyed late at night after reading sixty papers." The student is usually so grateful for the extra points that nothing more need be said.

To that (extremely) unusual student who says

> You are an incompetent boob. I am going to complain to the chairman of the department.

you could say "That is your privilege. Let me phone his/her secretary and make an appointment for you." Most of the time the student will back down. In the rare instances when the student does not retreat, any good chairman will give you ample opportunity to clarify the situation and smooth things over.

I don't mean to suggest in these pages that teaching is a confrontational activity. On the contrary, it can and should be a nurturing activity. But the potential for conflict is there, and I shall avail myself of several opportunities to suggest how you can either avoid or ameliorate conflict.

Another aspect of showing respect for your students is not springing nasty surprises on them. Do not say that you will grade a course one way and then grade it in a different fashion altogether. Do not act like an arrogant autocrat. (The truth is that, as a college teacher, you *are* an autocrat and a monarch and can do pretty much as you please. But there is no need to flaunt this before your students.) Do not create examinations that are full of dirty trick questions. If you do, you will be setting up a self-fulfilling situation in which everyone will do badly. Then you can brag to your pals that the students are stupid. How does this profit you? It is easier to make up a straightforward exam (not necessarily a watered-down exam) that tests students *on the material that you taught them.* If, as a result, the average on the test is 80% (a not very likely eventuality) then you can then tell the students how pleased you are that they have mastered the material so well.

If you give a miserable exam on which the average is 30% then you will have accomplished the following: for about 5% of the students (at the most) you will have set a standard to be risen to; but you will have alienated the other 95%. Students need not alienation but encouragement and (suitably tempered) challenge.

I have been known to complain to my colleagues that "students today do not know what it means to rise to a challenge". I was wrong to say this. It is true that today's students are not accustomed, as perhaps was a student like Bertrand Russell or G. H. Hardy, to have ever higher hurdles set before them at a prodigious rate. Many American students went through a grade school and a high school program in which they played a somewhat passive role. In a number of troubled high schools, students receive a grade of "A" or "B" provided only that they do not cause problems. This does not mean that students today are stupid. Rather, they may not have been challenged to rise to their full potential; they have never realized their capacity. In spite of television and other ostensibly insidious forces in our society, it is still a part of human nature to want to excel. It takes some practice to learn how to bring this out in students, but it can be done.

There are important philosophical and educational issues at play here. In America in the 1990s we attempt, as much as possible, to educate everyone. Whereas forty years ago this meant that "everyone" went to high school, now it means that "everyone" goes to college. It is vital in a free society that all citizens, regardless of financial resources, have an opportunity to pursue a college education. But society is set up in such a way now that a large percentage of young people go to college regardless of their interests or goals. For this we pay a price. We can rely less on the preparedness of our freshmen. We also can rely somewhat less on their attitudes and motivation.

What this means in practice is that, quite often, especially with freshmen—and especially at a public institution—we are not necessarily teaching a very select group. Many public institutions these days have an open admissions policy: anyone with a high school diploma has the right to attend the state university. From the taxpayers' point of view, such an admission policy makes perfectly good sense. If you are a professor at a state institution then you must make peace with the realities connected with such a policy. You must learn to adjust your expectations. You must learn a little patience, and learn to be flexible.

I am not about to recommend that college math teachers spend their evenings reading position papers on motivation such as are written at schools of education. I *am* recommending that the college mathematics teacher exercise some tolerance. Students *will* rise to a challenge, provided that the teacher starts with small challenges and works up to big ones. If students stumble at the first few challenges then they need encouragement, not derision. Exercising patience requires no more effort than exercising your vocal chords with an insulting remark.

The professor's attitude towards the class is apparent from his/her every word, every gesture, every action. If you are arrogant, if you despise your students, if you feel that you are above the task of teaching this course, then your students will get the message immediately. And what are you accomplishing by evincing these attitudes? Does it make you feel superior? More accomplished? More secure? More important? It should not. Proving a great theorem or writing a good book or article should make you feel secure and important and superior and accomplished. Doing a good job teaching the chain rule should make you feel as though you have done something worthwhile for someone else that day. There is something of value, of an intangible nature, about passing knowledge along to other people. Why not take some pleasure in it?

2. Attitude

I have long felt that those who cannot teach are those who do not care about teaching. If you actually care about transmitting knowledge then much of what I say in this booklet follows automatically. But some comments should be made.

Your students are a lot like you. When you enroll for a class, you have certain expectations. It is reasonable, therefore, that when you *teach* a class you should endeavor to live up to those same expectations. From this it follows that you should prepare, be organized, be fair, be receptive to questions, meet your office hour, and so forth.

On the other hand, your students are not like you. Especially in elementary courses, you cannot expect your students to be little mathematicians. Many of them are in the class *only* because it is a prerequisite for their major. Try to remember how you felt when you took anthropology or Latin or biology. Not everyone has a gift for mathematics. Unfortunately, some people have an attitude problem to boot (this attitude problem is sometimes termed "math anxiety"—see Section 1.13).

So you must learn to be sympathetic and receptive, and you must learn to be patient. Teaching is part of your craft, and part of your job. Perhaps if you are Gauss, or if you have just proved the Riemann hypothesis, then you can justifiably say that you are above these considerations. I'm betting that you are not either of these. If you call yourself a professor, and if you have the temerity to stand in front of an audience and profess, then you should show your audience some respect and consideration.

To stand in front of a class, with the charge of holding forth for an hour or more, is a heady experience—especially for the novice instructor. It is an ego trip. If you are prone to showing off anyway, this is an opportunity to let your predilections get out of hand. There is a temptation to tell too many jokes, to give a monologue, to use off-color language, to emphasize points with pratfalls or physical humor, to wear grotesque or offensive T-shirts or funny hats, to dress up like Isaac Newton, or to just be silly. A good rule of thumb is "Don't." There is a famous calculus teacher who used to wear a gorilla suit when he taught the chain rule. The idea was that the chain rule is so simple that even a monkey can do it. My view is that gimmicks such as this distract from the task at hand, which is to convey knowledge. How can a student concentrate on the mathematics if the instructor is dressed like a gorilla and acting silly to boot? If you want to wear a mask for the first couple of minutes of a lecture on Halloween, I guess that is all right. But do not introduce distractions into the classroom atmosphere.

For the length of a semester, you and your class are like a little family (if it is a large lecture, read "large, loosely knit, family"). The class develops its own *gestalt* and set of attitudes. Things will go smoothly if the attitude in your class is that you and the students are working together to conquer the material. If instead it is you and the book pitted against the students, then you've got an attitude problem. If, on the other hand, you take repeated pains to criticize the text, then you are setting up another attitude problem. No text is perfect; neither are the students and neither are you. But make it clear from the outset that you are on the students' side. You convey this attitude in thought, word, and deed: Prepare your lectures, respect student questions, give fair exams, meet your office hours.

Implicit in this discussion is a simple point: a successful class is not a confrontation between the professor and the students. The professor and the students should be allies, with the former playing the role of mentor, in mastering the material at hand. If that is not your role as teacher then what, pray tell, could it possibly be?

Your classes should be friendly, but you do not want to be friends with your students. This sounds a trifle cold, especially to a graduate student or new instructor, but it is an important device in maintaining control of the class. A slight distance helps preserve your authority. In particular, don't allow your students to call you "Bubbles" or any other affectionate name. It is probably not even a good idea to let your students address you by your first name.

Of course you should not date your students. It is safest not to date any student at your college or university. Given the way that students like to gossip about faculty, if you date one student you may as well be dating them all.

However, if you are inexorably smitten with one of your own students, then advise your love interest to transfer to another class or another section. Sexual harassment is an issue of great concern these days (Sections 3.7, 3.8). Professors are particularly vulnerable to charges of sexual harassment. Behave accordingly.

It is natural to want to show students your human side. There is probably no intrinsic harm in having a cup of coffee with some students. A relaxed atmosphere can help to open lines of communication. There might be some harm in having coffee with just one student—examine your conscience before doing this. If you meet a student in a bar off campus for a drink, then—let me speak frankly—you've got more on your mind than teaching.

Now let's return to teaching proper. A recurring theme in this booklet is that you should *prepare* your lectures. For a novice instructor, an hour or two or three of preparation may be necessary. For a seasoned trouper teaching calculus for the tenth time, as little as thirty minutes may be sufficient. The main thing is to be sure that you can do the calculations and that you have the definitions and theorems straight, and in the proper order. If you are the sort of person who freezes up in front of an audience, then be extra well-prepared. I know experienced professors who get so locked up in front of a large group that they cannot remember their own phone numbers. If that describes you, then have the necessary information on a sheet of paper.

If it is evident to your students that you are winging your lecture, then they are receiving a counterproductive message: if it is OK for the instructor to fake it then it is OK for the students to fake it. To those math teachers who say "I don't prepare because it is good for the students to see how a mathematician thinks" I say "nonsense." This is just laziness and/or self-serving rubbish. You must set a role model, both as an educated person and as to the way that mathematics is done.

The first few lectures that you give in a semester-long class set the tone for the entire semester. You may be at a slight disadvantage because you are coming off summer vacation. Perhaps you are not quite yet in the mood to be teaching, and your lack of enthusiasm shows. We've all fallen into the trap of saying, *sotto voce*, "I'll just wing the first few lectures and get things straightened out in a few weeks." This is a mistake. It is sending the wrong message to the class about your attitude towards discipline (both academic and personal), your attitude towards the students, and your attitude towards the subject matter. To repeat, you are a role model.

What your students write on their homework and on their exams will be a derivative version of what you show them. If you do not work out maximum-minimum problems systematically then they will not either. I always set up six steps to follow in doing a max-min problem and follow them scrupulously. This step-by-step approach is an elementary device, but it is an effective one for keeping interest up. When doing an example, after I do step two I can say "OK, so what do we do next" and this keeps the ball rolling.

Mathematicians fall unthinkingly into the use of jargon. Among ourselves, we frequently say 'trivially', 'clearly', 'by inspection', and so forth. Do not do this—especially in a calculus or pre-calculus class. First, it sounds pretentious. Second, it is dangerous to assume that anything is either trivial or clear unless you make it so. Third, to say that something is trivial is a subtle put-down: in the popular psycho babble this would be called "passive-aggressive behavior."

3. Personal Aspects

Like many activities in life, teaching is an intensely personal one. Some teachers have a lighthearted, informal, even jocular style. Others are more severe. Some give a rigid, structured lecture. Others conduct a Socratic interchange with the class. Some send students to the board to do problems (some, in the R. L. Moore style, do nothing but). Some instructors use overhead slides, computer simulations, symbolic manipulation software, and MATHEMATICA graphics. Others do it all themselves, with just a piece of chalk. Some professors integrate a (computer-based) laboratory component into their courses. [In fact I would like to see mathematics become more of a "laboratory discipline." But I defer a discussion of that topic to another time.] All of these methods are correct. It is essential for you to be comfortable with your class. Therefore you should conduct the class in whatever fashion feels most natural to you.

However you should be willing to try new things. If you have never told a joke before, try telling a joke. If it works, you may be pleasantly surprised and may tell another. [But be forewarned: eighteen year olds are insecure and are always worried that someone is making fun of them. Do not tell jokes that may be interpreted in that fashion. Do not tell jokes at anyone's expense. Do not tell sexist jokes. Do not use vulgar language.] Try introducing the product rule with a story about how much trouble Leibniz had getting it right. Illustrate the importance of the constant of integration by integrating $\int 1/x\,dx$ by parts (without the constant) and deriving the assertion that $0 = 1$:

$$\int \frac{1}{x}\,dx = \frac{1}{x} \cdot x - \int x \cdot \left(-\frac{1}{x^2}\right)\,dx = 1 + \int \frac{1}{x}\,dx$$

hence $0 = 1$. Some of these endeavors will fall flat. Others will breathe new life into an otherwise old and (for you) dull topic.

It is important to me that my classroom have the atmosphere of an interchange of ideas among intelligent people. I would be most uncomfortable to stand for an hour reciting a litany to a sea of blank faces. Thus I am continually trying new approaches, new angles, new ideas. It is a way to keep my lectures fresh, even in a course that I have taught ten times before.

I do not find it useful to send students to the blackboard to do problems. First, the time that it takes for the student to get to the front of the room, falter around, and sit down again, is too great for the benefit obtained. I do everything myself because I can teach a great deal even while I am doing the most mundane example. But others have been sending students to the board for years and swear by it. Do what works for you.

I have no use for overhead projectors. To me, part of pacing a lecture is letting it evolve on the blackboard. Part of the dynamic of my lecture is moving back and forth in front of the material. But others find that they can be more organized if they write out the material in advance on overhead slides. Still others write the material in real time on the overhead slide. Yet another group writes very little, but stands in one spot and delivers a strictly oral lecture.

Remember that you are delivering a product. Cadillac does this differently from Mercedes Benz. You must develop your own delivery. The over-riding consideration is that you be comfortable with your delivery so that you can make your class comfortable in turn. The style, organization, and content of your class is a reflection of you and your attitude towards the class. If you stand in front of your calculus class facing the blackboard, mumbling to yourself, and writing "Theorem–Proof–Theorem–Proof" then what message are you sending to the students? If instead you do a stand-up comedy routine and get around to the mathematics in the last ten minutes of class, then what message are you sending to the students?

Speaking strictly logically, it should not make any difference whether you wear a suit, or jeans and a work shirt, or wear a loin cloth and carry a spear when you teach. But it does. Dressing nicely sends a subtle signal to the class that you are the person in charge. Straightening your tie and combing your hair before going to class is like putting on your mortar board before going to graduation; you are pausing to say to yourself "now I am going off to do something important."

4. Prepare

Some people rationalize not teaching well by saying (either to themselves or to others) "My time is too valuable. I am not going to spend it preparing my calculus lecture. I am so smart that I can just walk into the classroom and wing it. And the students will benefit from watching a mathematician think on his feet." (As a student, I actually had professors who announced this to the class on a regular basis).

It is true that most of us can walk into the room most of the time and mostly wing it. But most of us will not do a very effective teaching job if we do so. Thirty minutes can be sufficient time for an experienced instructor to prepare a calculus lecture. A novice instructor, especially one teaching an unfamiliar subject for the first time, may need considerably more preparation time. Make sure that you have the definitions and theorems right. Read through the examples to make sure that there are no unpleasant surprises. It is a good idea to have a single page of notes containing the key points. To write out every word that you will say, write out a separate page of anticipated questions, have auxiliary pages of extra examples, have inspirational quotes drawn from the works of Thomas Carlyle, make up a new notational system, make up your own exotic examples, and so forth is primarily an exercise in self abuse. Over-preparation can actually stultify a lecture. But you've got to know your stuff.

I cannot over-emphasize the fact that preparation is of utmost importance if you are going to deliver an effective lecture or give a stimulating class. However it is also true that the more you prepare the more you lose your spontaneity. You must strike a balance between (i) knowing the material and (ii) being able to "talk things through" with your audience.

My own experience is that there is a "right amount" of preparation that is suitable for each type of course. I want to be confident that I'm not going to screw up in the middle of lecture; but I also want to be actually thinking the ideas through as I present them. I want to feel that my lecture has an edge. It *is* possible to over-prepare. To continue after you have prepared sufficiently is a bit like hitting yourself in the head with a hammer because it feels so good when you stop.

You must be sufficiently confident that you can field questions on the fly, can modify your lecture (again on the fly) to suit circumstances, can tolerate a diversion to address a point that has been raised. The ability to do this well is largely a product of experience. But you can *cultivate* this ability too. You cannot learn to play the piano by accident. And you will not learn to teach well by accident: you must be aware of what it is that you are trying to do and then consciously hone that skill.

If you do not prepare—I mean *really* do not prepare—and louse up two or three lectures in a row, then you will experience the following fallout: (i) students will take up your time after class and during your office hour (in order to complain and ask questions), (ii) students will complain to the undergraduate director and to the chairman, (iii) students will (if you are really bad) complain to the dean and write letters to the student newspaper, (iv) students will write bad teaching evaluations for your course.

Now student teaching evaluations are not gospel. They contain some remarks that are of value and some that are not. Getting bad teaching evaluations does not necessarily mean that you did a bad job. And I know that the dean will only slap me on the wrist if he gets a complaint about my teaching (however if there are ten complaints then I had better look out). Finally, I know that the chairman will give me the benefit of the doubt and allow me every opportunity to put any difficult situation in perspective. But if I spend thirty minutes preparing each of those three lectures then I will avoid all this grief and, in general, find the teaching experience pleasurable rather than painful. What could be simpler?

As well as preparing for a class, you would be wise to de-brief yourself after class. Ask yourself how it went. Were you sufficiently well prepared? Did you handle questions well? Did you present that difficult proof as clearly as you had hoped? Was there room for improvement? Be as tough on yourself as you would after any exercise that you genuinely care about—from playing the piano to a having a tennis match. It will result in real improvement in your teaching.

Read your teaching evaluations. Many are insipid. Others are puerile. But if ten of the students say that your writing is unclear, or that you talk too quickly, or that you are impatient with questions, then maybe there is a problem that you should address. Teaching is a yoga. Your mantra is "am I getting through to them?"

It is a good idea to try to anticipate questions that students will ask. But you cannot do this artificially, as a platonic exercise late at night over a cup of coffee. It comes with experience. Assuming that you have adopted the attitude that you actually care whether your students learn something, then after several years of teaching you will know by instinct what points are confusing and why. This instinct enables you to prepare a cogent lecture. It helps you to be receptive to student questions. It helps you to have a good attitude in the classroom.

An easy way to cut down on your preparation time for a lecture is to present examples straight out of the book. The weak students will appreciate this repetition. Most students will not, and you will probably be criticized for this policy. On the other hand, it is rather tricky to make up good examples of maximum-minimum problems or graphing problems or applications of Stokes's theorem. It can be time-consuming as well. A good rule of thumb is this: if you need more examples for your calculus class, pick up another calculus book and borrow some. Develop a file of examples that you can dip into each time you teach calculus. You will learn quickly that making up your own examples is hard work. Do you ever wonder why most calculus books are so disappointing? All right, *you* try to make up eight good examples to illustrate the divergence theorem.

5. Clarity

When you teach a mathematics class, clarity (or lack thereof) manifests itself in many forms. If you are the most brilliant, and even the most well-prepared, mathematics lecturer in the world, but you stand facing the blackboard and mumbling to yourself, then you are not being clear. If instead you shout at the top of your lungs so that all can hear, but if your handwriting is cryptic, then you are not being clear. If your voice is clear, your handwriting clear, but your blackboard technique non-existent, then you are not being clear. If your voice is beautiful, your handwriting artistic, your blackboard technique flawless, but you are completely disorganized, then you are not being clear. If you speak clearly, write clearly, have good blackboard technique, are well organized, but speak with a foreign accent, then don't worry. You are being clear.

Here is the point. Mathematics is hard. Do not make it harder by putting artificial barriers between you and your students. If you are shy and simply cannot face your audience, then perhaps you chose the wrong profession. More seriously, be very well prepared. *Make* yourself confident. Calculus is one of the most powerful analytic tools that has ever been created. It is a privilege to be able to pass it along to the next generation. Be proud of what you are doing. It is no less an event for you to teach the fundamental theorem of calculus to a group of freshmen in the 1990s than it was for Archimedes to teach his students how to calculate the area of a circle.

I have atrocious handwriting. When my departmental librarian got her first written message from me she thought it had been written in Chinese hieroglyphs. But when I lecture I slow down. I write deliberately. I *want* my audience to understand me and to respect me and I take steps to see that this actually happens.

Suppose that you are in the middle of a lecture and you are making a very important point. How can you drive it home? How can you get the students' attention? We all know that students drift into a malaise in which they are copying and not thinking (after all, we were once students and did the same). How do you wake them up? It's easy. Pause. State the point clearly and simply. Write it clearly and simply. Say "This is important." Repeat the point. One of Mozart's most effective tools in his compositions was to repeat a particularly beautiful passage. We can benefit from his example.

Ask whether there are any questions. Repeat the point again. Assure students that this point will be on the exam, and that it will come up over and over again in the course. Tell them that if they do not understand this point then it will hamper them later in the course. Knowing how to make certain that students know when you are making an important point is a big—and infrequently mentioned—aspect of the "clarity" issue.

If your teaching evaluations say that "the exam didn't cover the points stressed in class," it may mean that you don't know how to write a good exam (Section 2.9). But it may also mean that you don't know how to put your point across, or how to tell the students what is important—in other words, how to make yourself clear.

6. Speak up

If you are going to be a successful lecturer then you have to find a way to fill the room with yourself. If you stand in front of the class (be it a class of ten or a class of a hundred) and mumble to yourself then you will not successfully convey the information. Even the most dedicated students will have trouble paying attention. You will not have stimulated anyone to think critically.

You do not need to be a show-off or a ham or a joke teller to fill a room with your presence. You can be dignified and reserved and old-fashioned and still be a successful lecturer with today's students. But you must let the students know that you are there. You must establish eye contact. You must let them know that you are *talking to them.*

Before I start a lecture, especially to a large class, I engage some of the students in informal conversations. I get them to talk about themselves. I ask them how they are doing on the homework assignment. I comment about the weather. Then I make a smooth transition into the lecture. That way I already have half a dozen people on my side. The others soon follow.

Some new instructors—especially those who are naturally soft spoken or shy—may need some practice with voice modulation and projection. Get together a group of friends and give a practice lecture for them. Ask for their criticism. Make a tape recording of your practice lecture and listen to it critically.

If there is any doubt in your mind as to whether you are reaching your audience during a lecture, then *ask* about it. Say "Can you hear me? Am I talking loud enough? Are there any questions?" This is one of many simple devices for changing the pace of a lecture, giving note-takers a break, allowing students to wake up.

Think of a good movie that you've seen recently. Now remove the music; remove the changes in focal length; remove the changes of scenery; remove the voice modulation and changes of emotion; remove the skillful use of silence as a counterpoint to sound. What would remain? Could you stay awake during a showing of what is left of this movie? Now think about your class in these terms.

7. Lectures

In an empty room sits a violin.

One person walks in, picks it up, draws the bow across the strings, and a horrible screeching results. He leaves in bewilderment.

A second person walks in, attempts to play, and the notes are all off key.

A third player picks up to the instrument and produces heavenly sounds that bring tears to the eyes. He is Isaac Stern and the instrument is a Stradivarius.

Wouldn't it have been foolish to say, after hearing the first two players, that this instrument is outmoded, that it doesn't work? That it should be abandoned to the scrap heap? Yet this is what many are saying today about the method of teaching mathematics with lectures. Citing statistics that students are not learning calculus sufficiently well, or in sufficiently large numbers, government sponsored projects nationwide assert that the lecture doesn't work, that we need new teaching techniques.

Those who say that "the use of the lecture as an educational device is outmoded" rationalize their stance, at least in part, by noting that we are dealing with a generation raised on television and computers. They argue that today's students are too ready to fall into the passive mode when confronted with a television-like environment. It follows that we must teach them interactively, using computers and software to bring them to life.

Lectures have been used to good effect for more than 3000 years. I am hesitant to abandon them in favor of a technology (personal computers, videos) that has existed for just ten years. In spite of popular rumors to the contrary, a lecture does not need to be a bone dry desultory Philippic. It can have wit, erudition, and sparkle. It can arouse curiosity, inform, and amuse. It is an effective teaching device that has stood the test of time. The ability to give a good lecture is a valuable art, and one that you should cultivate.

However you really have to work at making your lectures reach your students. It is true that mathematics teaching in this country is not, overall, very effective. The reason, however, is not that the lecture method is "broken." Rather, we tend not to put a lot of effort into our teaching because the reward system is often not set up to encourage putting a lot of effort into it. You must learn to develop eye contact with your audience, to fill the room with your voice and your presence, and to present your ideas with enthusiasm and clarity. Other sections of this booklet deal in detail with these techniques.

Turn on your television and watch a self-help program, or a television evangelist, or a get-rich-quick real estate huckster. These people are not using overhead projectors, or computer simulations, or MATHEMATICA. In their own way they are lecturing, and *very effectively*. They can convince people to donate money, to change religions, or to join their cause. Of course your calculus lecture should not literally emulate the methods of any of these television personalities. But these people and their methods are living proof that the lecture is not dead, and that the traditional techniques of Aristotelian rhetoric are as effective as ever.

There are other useful teaching environments besides lectures. Although less common in mathematics than in some humanities courses, group discussions can be useful. If you want to get students interested in what the boundary of a set in a metric space ought to be, begin with a discussion in which students offer various suggestions. Before you define what a finite set is, ask the students to suggest a definition. See also Section 2.5 on the benefits of group activity.

As already mentioned, it can also be instructive to have students volunteer to do problems at the blackboard. Once in a great while, if a student offers an alternate proof of a proposition or another point of view, I hand him/her the chalk. Everyone is usually quite surprised, but the results are generally pleasing and it provides a nice change of pace.

Computer labs can also be a useful instructional device. The subject of sophomore level differential equations lends itself well to helping students explore the interface between what we can do by hand and what the machine can teach us. Let's be frank: we do not know how to solve most differential equations. Thus it is important for students to see how much analysis one can do with traditional methodologies and then to see how the machine can use phase plane analysis, numerical methods, and graphing to provide further concrete data.

It is important that we constantly seek new and better methods and technologies for teaching. This author, and this booklet, has a built in bias toward traditional methods, such as lectures. That is because he has watched them work and used them successfully for more than twenty years. I hope that other writings will explore some of the new teaching techniques.

8. Questions

In a programmed learning environment, whether the interface is with a PC or with MATHEMATICA notebooks or with a MAC, the student cannot ask questions. The give and take of questions and answers is a critical aspect of the human part of the teaching process. Teachers are *supposed* to answer questions.

There is more to this than meets the eye. When I say that a teacher answers questions I do not envision the student saying "What is the area of a circle?" and the teacher saying " πr^2." I instead envision the student struggling to articulate some confusion and the experienced teacher turning this angst into a cogent question and then answering it. To do this well requires experience and practice. I frequently find myself responding to a student by saying "let's set your question aside for a minute and consider the following." I then put the student at ease by quickly running through something that I know the student knows cold, and that serves as a setup for answering the original question. With the student on my side, I can answer the primary problem successfully. The point is that some questions are so ill-posed that they literally cannot be answered. It is the teacher's job to make the question an answerable one and then to answer it. See also Section 2.19 on asking and answering questions.

A similar, but alternative scenario is one in which the student asks a rather garbled question and I respond by saying "Let me play the question back for you in my own words and then try to answer it ... ." The point is that the responses "Your question makes no sense" or "I don't know what you mean" are both insulting and a cop-out. To be sure, it is the easy answer; but you will pay for it later. It takes some courage for the student to ask a question in class; by treating questions with respect, you are both acknowledging this fact and helping someone to learn.

Yet another encouraging response to a student question is to say "Thank you. That question leads naturally to our next topic ... " Of course you must be quick on your feet in order to be able to pull this off. It is worth the trouble: students respond well when they are treated as equals.

There are complex issues involved here. A teacher does not just lecture and answer questions. A good teacher helps students to discover the ideas. There are few things more stimulating and rewarding than a class in which the students are anticipating the ideas because of seeds that you have planted. The way that you construct your lecture and your course is one device for planting those seeds. The way that you answer questions is another.

When I discuss teaching with a colleague who has become thoroughly disenchanted with the process, I frequently hear complaints of the following sort: "Students these days are impossible. The questions that they pose are unanswerable. Suppose, for example, that I am doing a problem with three components. I end up writing certain fractions with the number 3 in the denominator. Some student will ask 'Do we always put a 3 in the denominator when doing a problem from this section?' How am I supposed to answer a question like that?"

Agreed, it is not obvious how to answer such a question, since the person asking it either (i) has not understood the discussion, (ii) has not been listening, or (iii) has no aptitude for the subject matter. It is tempting to vent your spleen against the student asking such a question. Do not do so. The student asking this question probably needs some real help with analytical thinking, and you cannot give the required private tutorial in the middle of a class hour. But you can provide guidance. Say something like "When a problem has three components it is logical that factors of 1/3 will come up. This can happen with certain problems in this section, or in any section. But it would be wrong to make generalizations and to say that this is what we do in all problems. If you would like to discuss this further, please see me after class." In a way, you are making the best of a bad situation; but at least you are doing something constructive, and providing an avenue for further help if the student needs it.

Let us consider some other illustrations of the principle of making a silk purse from a sow's ear:

The first example is a simple one.

Q: Why isn't the product rule $(f \cdot g)' = f' \cdot g'$?

The answer is *not* "Here is the correct statement of the product rule and here is the proof." Consider instead how much more receptive students will be to this:

A: Leibniz, one of the fathers of calculus, thought for many years that this is what the product rule should be. But he was unable to verify it. Of course Leibniz was hampered because he didn't have the language of functions. We do. If we set $f(x) = x^2$ and $g(x) = x$ then we can see rather quickly that $(f \cdot g)'$ and $f' \cdot g'$ are unequal. So the simple answer to your question is that the product rule that you suggest gives the wrong answer. Instead, the rule $(f \cdot g)' = f' \cdot g + g' \cdot f$ gives the *right* answer and can be verified mathematically.

The second example is more subtle.

Q: Why don't we divide vectors in three-space?

The *wrong* answer is to tell about Stiefel-Whitney classes and that the only Euclidean spaces with a division ring structure are $\mathbb{R}^1, \mathbb{R}^2, \mathbb{R}^4$, and $\mathbb{R}^8$. A better answer is as follows:

A: J. Willard Gibb invented vectors to model physical forces. There is no sensible physical interpretation of "division" of physical forces. The nearest thing would be the operations of projection and cross product, which we will learn about later.

Notice that in both illustrations an attempt is made to turn the question into more than what it is—to make the questioner feel that he/she has made a contribution to the discussion.

Q: Why isn't the concept of velocity in two and three dimensions a number, just like it is in one dimension?

If you are in a bad mood, you will be tempted to think that this person has been dreaming for the last hour and has understood absolutely nothing that you have been saying. Bear up. Resist the temptation to voice your frustrations. Instead try this:

A: Let me rephrase your question. Instead let's ask "Why don't we use vectors in one dimension to represent velocity just as we do in two and three dimensions?"

One of the most important features of vector language is that a vector has *direction* as well as magnitude. In one dimension there are only two directions: right and left. We can represent those two directions rather easily with either a plus or a minus sign. Thus positive velocity represents motion from left to right and negative velocity represents motion from right to left. The vector language is *implicit* in the way that we do calculus in one dimension, but we need not articulate it because positivity and negativity are adequate to express the directions of motion.

In dimensions two and higher there are infinitely many different directions and we require the explicit use of vectors to express velocity.

As the author of this booklet, I have the luxury of being able to sit back and think carefully about how to formulate these "ideal" answers to poor questions. When you are teaching you must be able to do this on your feet, either during your office hour or in front of a class. At first you will not be so articulate. This is an acquired skill. But it is one *worth acquiring.* It is a device for showing respect for your audience, and in turn winning its respect.

A final note about questions. Even though you are an authority in your field, there are things that you don't know. Occasionally these lacunae in your knowledge will be showcased by a question asked in class or during your office hour (it does not happen often, so don't get chills). The sure and important attribute of an intelligent, educated individual is an ability to say "I don't know the answer to that question. Let me think about it and tell you next time." On the (rare) occasions when you have to say this, be sure to follow through. If the item that you don't know is an integral part of the class—and this had better not be the case very often—get it down cold because the question is liable to come up again in a different guise later in the course. If it is not an integral part of the course, then you have no reason to feel inadequate. Just get it straight and report back.

The main point here is this: do not under any circumstances try to fake it. If you do, you will look bad, your interlocutor will be frustrated and annoyed, and you will have served no good purpose. If there is any circumstance in which honesty is the best policy, this is it. Professor of Economic History Jonathan R. T. Hughes was wise to observe that "There is no substitute for knowing what you are talking about."

9. Inductive vs. Deductive Method

It is of paramount importance, epistemologically speaking, for us as scholars to know that mathematics can be developed *deductively* from certain axioms. The axiomatic method of Euclid and Occam's Razor have been the blueprint for the foundations of our subject. Russell and Whitehead's *Principia Mathematica* is a milestone in human thought, although one that is perhaps best left unread. Hilbert and Bourbaki, among others, also helped to lay the foundations that assure us that what we do is logically consistent.

However mathematics, as well as most other subjects, is not learned deductively: it is learned *inductively.* We learn by beginning with simple examples and working from them to general principles. Even when you give a colloquium lecture to seasoned mathematicians, you should motivate your ideas with good examples. The principle applies even more assuredly to classes of freshmen and sophomores.

Take the fact that the mixed partial derivatives of a C^2 function in the plane commute. To state this theorem cold and prove it—before an audience of freshmen—is showing a complete lack of sensitivity to your listeners. Instead, you should work a couple of examples and then say "Notice that it does not seem to matter in which order we calculate the derivative. In fact there is a general principle at work here." Then you state the theorem.

Whether you actually give a proof is a matter of personal taste. With freshmen I would not. I'd tell them that when they take a course in real analysis they can worry about niceties like this. Other math instructors may opt for the more rigorous approach.

And by the way—you know and I know that C^2 is too strong a hypothesis for the commutation of derivatives. But really, isn't that good enough for freshmen? If a bright student raises this issue, offer to explain it after class. But do not fall into the trap of always stating the sharpest form of any given result. Great simplifications can result from the introduction of slightly stronger hypotheses, and you will reach a much broader cross-section of your audience by using this device.

Now suppose that you are teaching real analysis (from [RUD], for example). One of the neat results in such a course is that a conditionally convergent series can be rearranged to sum to any real limit. When I present this result, I first consider the series $\sum(-1)^j/j$ and run through the proof specifically for this example. The point is that, by specializing down to an example, I don't have to worry about proving first that the sum of the positive terms diverges and the sum of the negative terms diverges. That is self-evident in the example. Thus, on the first pass, I can concentrate on the main point of the proof and finesse the details.

Go from the simple to the complex—not the other way. It's an obvious point, but it works. Here is another example of that philosophy, implemented somewhat differently. Many calculus books, when they formulate Green's theorem, go to great pains to introduce the notions of "x- simple domain" and "y- simple domain" (i.e. domains with either simple horizontal or simple vertical cross-sections). This is because the authors are looking ahead to the proof, and want to state the theorem in precisely the form in which it will be proved. The entire approach is silly.

Why not state Green's theorem in complete generality? Then it is simple, sweet, and students can see what the principal idea is. When it is time for the proof, just say "to keep the proof simple, and to avoid technical details, we restrict attention to a special class of domains ... " This approach communicates exactly the points that you wish to convey, but cuts directly to the key ideas and will reach more of the students with less fuss.

Here is a useful device—almost never seen in texts or discussed in teaching guides—that was suggested to me by Paul Halmos:

Suppose that you are teaching the fundamental theorem of algebra. It's a simple theorem; you could just state it cold and let the students think about it. But the point is that these are ***students***, *not* mathematicians. It is your job to give them some help. First present to them the polynomial equation $x - 7 = 0$. Point out that it is easy to find all the roots and to say what they are. Next treat $2x - 7 = 0$. Follow this by $x^2 + 2x - 7$ (complete the square—imitating the proof of the quadratic formula). Give an argument that $x^3 + x^2 + 2x - 7$ has a root by using the intermediate value property. Work a little harder to prove that $x^4 + x^3 + x^2 + 2x - 7$ has a root. Then surprise them with the assertion that there is no formula, using only elementary algebraic operations, for solving polynomial equations of degree 5 or greater. Finally, point out that the remarkable fundamental theorem of algebra, due to Gauss, guarantees in complete generality that any non-constant polynomial has a (complex) root.

Notice how much depth and texture this simple discussion lends to the fundamental theorem. You have really given the students something to think about. Stating the theorem cold and then moving ahead, while *prima facie* logical and adequate, does not constitute teaching—that is, it does not contribute to understanding. As with many of the devices presented in this booklet, this one becomes natural after some practice and experience. At the beginning it will require some effort. The easiest thing in the world for a mathematician to do is to state theorems and to prove them. It requires more effort to *teach*.

Beresford Parlett recently said

> Only wimps do the general case. Real teachers tackle examples.

I think that Halmos's ideas illustrate what Parlett is saying.

One could go on at length about the philosophy being promulgated here. But the point has been made. Saki once said that "A small inaccuracy can save hours of explanation." Mathematicians cannot afford to be inaccurate. But, for the students' sake, we can simplify.

10. Advanced Courses

The teaching problems that arise in an advanced course are rather different from those in a lower division course. You are dealing with a more mature audience and, in at least some advanced courses, many of your students will be math majors. The main message for a new teacher is: don't get carried away. Don't try to tell them about your Ph.D. thesis in the first week of class. Try to remember the troubles you had learning about uniform continuity and uniform convergence. Give lots of examples. Prepare your lectures well and *slow down*. Be receptive to questions and sympathetic to awkward struggles with sophisticated new ideas. Be willing to repeat yourself.

It is probably best in an advanced undergraduate course to cover less material but to cover it in depth—to endeavor to give the students a real feel for the subject—rather than to race through a lot of material. Again I shall repeat an implicit theme of this booklet: most undergraduate students do not have either the maturity or the experience to put the shortcomings of your teaching into perspective. Good teaching is your responsibility.

Even in these courses, be sure to use the inductive method in favor of the deductive method as a vehicle for conveying ideas (refer to Section 1.9). Any hard theorem should be suitably motivated. Do even more examples than seems necessary. Refer to the Beresford Parlett quote at the end of Section 1.9 for inspiration.

It is tempting in an upper division course to assume, at least subliminally, that your students are little mathematicians. They are not. This course may be their first exposure to rigorous thinking, to ϵ's and δ's, to Theorem–Proof–Theorem–Proof, to careful use of "for all" and "there exists", to quashing a possible theorem with a single counterexample. In short, you are not just teaching these students some advanced mathematics; you are also teaching them how to think. This is an important opportunity for you, the instructor; and it is an important juncture in the students' education. You must use it wisely.

Today's undergraduate students do not have the background and experience in rigorous thinking that we all fancy we had when we were students. They are unaccustomed to proofs and to the strict rules of logic. It is often a good idea to have a whirlwind review of logic at the beginning of an upper division mathematics course—especially in real analysis or algebra where *modus ponendo ponens*, contrapositive, proof by contradiction, induction, and so forth are used frequently. It would not be out of place to present some material on set theory and number systems as well. Some mathematics departments have a "transitions" course designed to bridge the gap in methodology between lower division courses and upper division courses. If yours is such a department, then you may moderate the advice in this paragraph.

If you do not make some extra effort to help the students in your advanced courses over the "hump" that separates math enthusiasts from mathematicians, then you are missing an opportunity to contribute to the pool of mathematical talent in this country. It's your decision, but if you decide not to participate then you have no right to complain when your department's graduate program is reduced due to lack of students, or your undergraduate program curtailed for lack of majors.

11. Time

There are several aspects of teaching that require time management skills. When you are giving a lecture, you must cover a certain amount of material in the allotted time—and at a reasonable rate. When you give a course, you must cover a certain amount of material in one semester. When you give an exam, it must be do-able by an average student in the given time slot. When you answer a question, the length of the answer should suit the occasion.

All of these topics, save the last, have been touched upon in other parts of this booklet. They require some thought, and some practice and experience, so that they become second nature to you.

Nobody can design a lecture so that the last 'QED' is being written on the blackboard just as the bell rings. There are certain precepts to follow in this regard:

- Have some extra material prepared to fill up extra time.
- If you finish your lecture with five minutes to spare, don't rocket into a new topic. You will have to repeat it all next time anyway, and students

find this practice confusing.

- If the clock shows that just five minutes remain, and you have ten or fifteen minutes of material left to present, then you will have to find a comfortable place to quit. Don't race to fit all the material into the remaining time. If possible, don't just stop abruptly, thinking that you can pick up a calculation cold in the following lecture.
  An experienced lecturer will know which will be the last example or topic in the hour, and that he/she might get caught for time. Therefore the lecturer will plan in advance for this eventuality and think of several graceful junctures at which he/she might bring the hour to a close. With enough experience, you will know intuitively how to identify the comfortable places to stop; thus end-of-the-hour time management problems can be handled on the fly. In particular, if five minutes remain then *do not begin* a ten minute example!
- If you prepare (the last part of) your lecture in units of five minutes duration, and if you are on the ball, then you should never have to run over by more than two minutes nor finish more than two minutes ahead of time. (The idea here is if there are three minutes remaining then you can include another five minute chunk without running over by more than two minutes. If there are just two minutes remaining then you should stop.)
- If you *run out of time*, do not run the class over the hour—at least not by more than a minute or two. Students have other classes to attend, and they will not be listening. If the time is gone, then just quit. Make up for your lapse in the next class (this will require some careful planning on your part). Best is to plan your lecture so that you do not fall behind.
  A special note about buzzers: Some math buildings have a loud buzzer or bell that sounds at the end of the hour. Once that buzzer sounds, all is lost. Most students will instantly start packing up their books and heading for the exit. At a school without a buzzer (especially one without a clock on the wall!) you have a bit of slack since no two wrist watches are in agreement. You may want to interpret the advice in this section according to the physical environment in which you are teaching.
- If a student asks a question that requires a long answer, don't let your answer eat up valuable class time. Tell the student that the question is ancillary to the main subject matter of the course (it had better be, or else you evidently forgot to cover an important topic) and that the question can best be treated after class. However do not let the student get the impression that the question is being given the brush off.
- Conversely, if a student asks a question for which a brief answer is appropriate (such as "shouldn't that 2 be a 3?" or "when is the next homework assignment due?") then do give a suitably brief answer. Anecdotes about your childhood in Shropshire are probably out of place.

By the way, this last is more than a frivolous remark. As we slide into our golden years, we seem to be irretrievably moved to share with our students various remembrances of things past: "it seems to me that twenty years ago students worked much harder than you people are willing to work" or "when I was a student, we put in 5 hours of study for each hour of class time" or "I used to walk six miles barefoot through the snow to attend calculus class". Trust me: students hate this. You will defeat all the other good things that you do by giving in to this temptation.

If you have the time problem under control at the level of lectures, then you will have the ability to pace your course in the large as well. You should have a good idea how much material you want to cover. And when you plan the course you should allot a certain number of lectures for each topic. If you are teaching undergraduates, then they depend on your course for learning a certain body of material (that may be prerequisite for a later course). Don't short-change them.

A test should be designed for the allotted time slot. You can rationalize giving a two hour exam in a one hour time slot by saying to yourself that there is so much material in the course that you simply *had* to make the test this long. This is nonsense. The point of the exam is not to actually test the students on every single point in the course, but to make the students *think* that they are being tested on every point in the course. Ideally, the students will study everything; but your test amounts to a spot check. Even if you had a four hour time slot in which to give the exam, you couldn't really test them on everything, now could you?

If you give a two-hour exam in a one hour time slot, then you run several risks: that students will become angry, demoralized, alienated, or all three. Telling the students not to worry about their grade of 37/100 because the average was 32/100 does not work. Students are unable to put such information into perspective.

12. Why do we Need Mathematics Teachers?

I frequently ask myself why I am necessary. Can't a student just pick up a book (or boot up a piece of software) and learn calculus or any other basic subject? Instead of charging students $10,000 to $20,000 per year to attend a university, why don't we charge them somewhat less for admission to a good library? Why don't we just sell them a box of diskettes with a computerized French course, a computerized calculus course, and so forth?

Like all topics in this book, the present one is simple and has a fairly obvious answer. But this answer needs to be articulated.

Many students read their texts with little or no understanding. They see the words but they do not understand the concepts. They need someone to tell them what is important, to give priority to the ideas, to demonstrate the techniques, to respond to their questions. This is something that a computer, or even a book, will never be able to do. When a student comes to my office to ask me a question, I can listen to the question and know at what level to pitch the answer. After I've delivered that answer I can look at the student's face and tell whether he/she has understood. A computer or a book will never be able to interact with students in this fashion.

Put a different way, the college or university mathematics teacher has (or at least ought to have) the distinction of knowing his/her entire subject. Such an instructor can put any question into context and add perspectives that we could never (at least in the foreseeable future) expect from a piece of hardware.

In a larger sense, I can make adjustments in the way that I am teaching a class depending on how well the students are absorbing the ideas. I can provide stimulating material both for the average student and for the gifted student. In short, I can interact with the students *as people,* and at their level. That is the role of the teacher; it is a role that will never be supplanted by books or computers or other inanimate objects.

It has been said that an education is the product of the interaction of two first class minds. There is truth in this, for a bright and eager student asks questions, processes the answers, and asks more. Conversely, a good teacher anticipates questions, plants the seeds of new questions, and reaps the harvest.

If you want to be an effective teacher, you should give some thought to the points being made in this section, and how you can implement them in your classroom. If you are no more effective than a book or a diskette, then you are not doing your job.

13. Math Anxiety

About fifteen years ago the concept of "Math Anxiety" was invented, probably in a school of education. We don't hear much about math anxiety in math departments because such departments are full of people who don't have it. Math anxiety is supposed to be an inability by an otherwise intelligent person to cope with quantification and, more generally, with mathematics. Classic examples of math anxiety are the successful business person who cannot calculate a tip, or the brilliant musician who cannot balance a checkbook.

I have trouble balancing my checkbook too. That is in part why I now keep track of my checking account using accounting software. But I don't think that I have math anxiety. Nor do I think I have checkbook anxiety. I'm just careless.

We could enlarge the issue and wonder whether some people have speaking anxiety or spelling anxiety. Nobody ever discusses these maladies. Does that mean that they don't exist?

What sets mathematics apart is that it is unforgiving. Most people are not talented speakers or conversationalists, but comfort themselves with the notion that at least they can get their ideas across. Many people cannot spell, but rationalize that the reader can figure out what was meant (or else they rely on a spell-checker). But if you are doing a math problem and it is not right then it is wrong. Period.

Learning elementary mathematics is about as difficult as learning to play Malegueña on the guitar. But there is terrific peer support for learning to play the guitar well. There is precious little such support for learning mathematics. If the student also has a mathematics teacher who is a dreary old poop and if the textbook is unreadable, then a comfortable cop-out is for the student to say that he has math anxiety. His friends won't challenge him on this assertion; in fact they may be empathetic.

The literature, in psychology and education journals, on math anxiety is copious. The better articles are careful to separate math anxiety from general anxiety and from "math avoidance." Some subjects who claimed to have math anxiety have been treated successfully with a combination of relaxation techniques and remedial mathematics review.

I don't think that it is healthy for a mathematics teacher to worry about math anxiety. Your job is to teach mathematics. Go do it.

14. How do Students Learn?

Psychologists, sociologists, and anthropologists have been debating this question for decades. There is no general agreement on how students learn.

It is my opinion that the very best students tend to teach themselves. The instructor points out guide posts for such a student, and then the student's native intellect takes over.

On the other hand, weak students are often quite dependent on the instructor and (perhaps) the text and the lectures. If you agree that these people are worth teaching at all, then you must be there for them. Provide good lectures and a reasonable text for them to work with. Answer their questions. Be as helpful, and as encouraging, as possible to students who have the courage to come to your office for help.

Students of middling abilities are perhaps in the majority, and they share properties with both the best students and the worst. Know with certainty that you cannot please everyone.

But also strive at least to provide some stimulation for students of all abilities. The ability to do so, without an unreasonable amount of effort on your part, can come only with experience and determination. Even if you had a class with just three students, it is likely that their levels of ability would fall into two or more disparate categories. Thus there is no escaping the realities of having an uneven audience. Being forewarned, and being thoughtful, can help you to present your course so that you teach something of value to most of the students most of the time.

Let me cast these matters in a different light. An instructor who is demanding and difficult and who omits many details from his lectures will challenge the talented students and force them to go off and learn the material on their own. Certainly that was the nature of the graduate program that I attended, and it did me a world of good. On the other hand, an instructor who is lucid and who proceeds at a comfortable pace makes everything look too easy and can lull students into a false sense of security. That instructor will also bore the gifted students. How do you address both of these phenomena in your classroom?

Because at most universities we are training a widely diverse group of students, the issues raised in the last paragraphs are unavoidable. One solution, in a fundamental course like calculus for example, is to have several tracks: a calculus course for scientists and engineers, one for pre-medical students, and one for business students. This is a commonly used method to slice up student abilities so that the spectrum in any given classroom is not so broad. Another approach to the issue of widely varying student abilities might be to rethink the traditional classroom/lecture format and divide students into groups in which they can seek their own level. The final word has not been said on how to deal with these problems. See Section 2.5 for more on the value of group work.

It is important that we continue to explore new methods to teach mathematics. It has been cogently argued that the failure rate in the large lecture system is unacceptably high; that the retention rate is unacceptably low; that America is falling behind in the technology race because we are not training our young people effectively. However, in my view it is simplistic to lay the blame for this problem on the *formal method* by which we have been teaching mathematics. No methodology is perfect.

Let us own up to the fact that many of us—especially those trained at high-powered research departments—are often not trained to care about teaching. Many of us do not. That is not the way the value/reward system is set up. The traditional methods of teaching still have much to offer, provided that they are being used by people who are properly trained and who care.

A good resource for new ideas on the teaching of mathematics is the periodical *UME Trends*. It has one or more regular columns in each issue that discuss specific teaching techniques (such as the device of mathematical POST-IT notes—see Section 2.19).

15. Computers

Computers are everywhere, especially in mathematics departments. Software for teaching mathematics is also everywhere. Most publishers of basic mathematics texts are absolutely convinced that they cannot market their product without making extensive software resources available to mathematics instructors. It is not clear, as of this writing, that much of this software is actually being used in the classroom. This is in part because most instructors are not conversant with what is available, are not comfortable with these products, or simply cannot be bothered.

Another form of software for the classroom that is being heavily touted these days is the MATHEMATICA notebook. Briefly, these are self-contained environments by means of which students can interact with the computer over various mathematical issues *without* knowing anything about computing or computer languages.

All of this hardware and software raises fundamental issues about the way that mathematics is taught and the way that it ought to be learned. I drive my car every day and am perfectly comfortable in doing so while not knowing in intimate detail how it works. Ditto for my computer, my telephone, my television, and so forth. Some would say that a measure of how civilized we are is how many black boxes we are willing to use unquestioningly. To what extent should this point of view be allowed in the mathematics classroom?

Most of us were trained with the idea that the whole point of mathematics is to understand precisely why things work. To make the point more strongly, this attitude is what sets us apart from laboratory scientists. We make no statement unless we can prove it. We use no technique unless we fully grasp its inner workings. We would not dream of using the quadratic formula unless we knew where it came from. We would not be comfortable using the fundamental theorem of calculus unless we had seen its proof.

Yet in many colleges these days the linear programming course consists primarily of learning to use LINDO or any number of other canned packages for applying the simplex method and its variants. The statistics course consists primarily in learning to use SAS. The undergraduate numerical analysis course consists in learning to use IMSL.

It has been argued that the use of programmable calculators and MATHEMATICA notebooks in the calculus classroom will free students from the drudgery of calculation and will allow us to teach them how to analyze multi-step word problems (see, for instance, [STE]). Thus, it is hoped, our lower division calculus classes will be more closely tied to the way that mathematics is used in the real world. I see this as double-talk.

It is certainly the case that the less able students are so hampered by their inability to take the derivative and set it equal to zero or to find the roots of a quadratic equation or to calculate the partial fraction decomposition of a rational expression that they have little hope of successfully analyzing a multi-step word problem whose solution includes one or more of these techniques. But there is no evidence to support the (apparent) contention that a person who is unable to use the quadratic formula will somehow, if these technical difficulties are handled for him/her by a machine, be able instead to analyze conceptual problems. Using the quadratic formula is easy. Analyzing word problems is hard. A person who cannot do the first will also probably not be able to do the second—with or without the aid of a machine.

I think, as I have already indicated, that the computer can be used successfully to provide helpful graphical analyses. If you want to draw a surface in space and then move it around to analyze it from all sides, then there is nothing to beat MATHEMATICA. If you want to illustrate Newton's method, or use the Runge-Kutta method, or implement Simpson's rule, then a computer is almost indispensable. But the use of the computer should be based on a firm foundation of conceptual and technical understanding.

Some of the federally funded calculus projects have been teaching the entire calculus sequence from MATHEMATICA notebooks. The principal investigators in these projects claim that they can take students with no particular interest in mathematics, who have traditionally done poorly in their mathematics courses, and ignite in them a spark for the subject using these notebooks. They claim that the students get truly excited from interacting with the machine. And students who have never before received a good math grade before end up getting an "A" in the course. However follow-ups have been done on these students and it has been determined that the majority of them have no comprehension of the subject matter and little retention. When confronted with this information, one principal investigator has said "I think that the concept of 'understanding' has traditionally been over-emphasized."

What does this *mean*? If we are not trying to endow our students with understanding then just what are we doing? I think that it is perfectly all right for a trade school or a commercial business college to teach linear programming by training its students to use LINDO. But in a university mathematics department a student of linear programming should be taught the simplex method and why it works.

It is probably the case that students learning calculus from MATHEMATICA notebooks respond positively to the extra attention they are receiving, to the novelty of the teaching environment, and to the fact that making a mistake when interacting with the computer is less heinous than asking a dumb question in class. In the MATHEMATICA notebook environment, the student is probably more willing to try things and to experiment. We should think carefully about how to capitalize on the special features that computers can contribute to the learning experience. However it is clear that we do not yet know all the answers, nor have we realized all the potential.

Consider this: in my view, a student is better off spending an hour with a pencil—graphing functions just as you and I learned—than generating fifty graphs on a computer screen in the same time period. From first hand experience, I am absolutely sure what the first exercise will teach the student. It is not at all clear what the student gains from the second.

A wise man once told me that the computer is a solution looking for a problem. It is obviously a powerful tool in the right circumstances. The generation of minimal surfaces of arbitrary genus by Hoffman, Hoffman, and Meeks ([HOF]) is a dazzling use of computers; the simulations that these three mathematicians performed would have been inconceivable without this advanced computer graphics technology. MATHEMATICA is a powerful tool in the hands of the right user, as are AXIOM, MACSYMA and MAPLE. However it is not apparent that we have yet found the right use for the computer in the calculus classroom. There is no data to support the contention that the use of computers, or even calculators, in the classroom leads to a more effective *long term* learning experience.

Calculus is perhaps the most powerful body of analytic tools ever devised. All young scientists should learn calculus in essentially the traditional fashion so that they have these tools at their disposal. *Graphing a function* is one of the most basic processes of analytical thinking—analogous to finger exercises for the piano. The partial fractions technique is also one of the most far-reaching algebraic devices that we have. Integration by parts is perhaps the most ubiquitous and powerful tool in all of mathematical analysis. Letting a computer do these processes for the students abrogates much of what we have learned in the last three hundred years.

Let me assure you that I have discussed this point with the presidents of large high tech corporations and they agree with me absolutely. Use of the new technology should be layered atop a traditional foundation. That is what works in the classroom and that is what works in the real world.

16. Applications

One of the most chilling things that can happen to an unprepared, unseasoned faculty member is to have a belligerent student raise his/her hand and say "What is all this stuff good for?" And one of the most irresponsible things that a faculty member can say in response is "I don't know. That is not my problem." If you do not have an answer for this student question then you are not doing your job.

I have found it useful in all of my undergraduate classes to tell the students about applications of the techniques being presented *before* the aforesaid chilling question ever comes up. This requires a little imagination. If I am lecturing about matrix theory then I tell the students a little bit about image processing and image compression. Or I tell them about eigenvalue asymptotics for clamped beams and applications to the building of a space station (not coincidentally, this is a problem that I have worked on in my research). If I am lecturing about surfaces then I tell them about the many applications of surface design problems. If I am lecturing about uniform continuity or uniform convergence then I tell them about some of the applications of Fourier analysis. The pedagogical technique that I am describing in effect defuses any potential belligerence from engineering or other students who have no patience for mathematical abstraction.

Carrying out this teaching technique requires a little forethought and a little practice. After a while it becomes second nature, and you will find yourself thinking of potential answers while taking a shower or walking to class. If it suits your style, keep a file of clever applications of elementary mathematics. It is not true that the concept of uniform convergence is used on a daily basis by engineers to construct bridges. Do *not* use this facile line of reasoning to talk yourself into abandoning the effort to acquaint lower division students with the applications of mathematics. Instead, reason that uniform convergence is a bulwark of the theory behind the practical applications of mathematics. It is important. Act as though you believe it.

Try to be flexible and to reach out for up-to-date and striking applications. Uniform convergence is a basic idea in the convergence of series; one of the most interesting uses of series is in Fourier analysis. And what is Fourier analysis good for? Mention the hot new theory of wavelets and some of its uses. Go from there.

It is a matter of personal taste (and much debate) as to how much should be done with applications in, say, a calculus class. For most of us the problem is solved by the very nature of the undergraduate curriculum. There is little time to do any but the most routine applications. But times are changing and many mathematics departments are re-evaluating their curricula. There is considerable enthusiasm for infusing the freshman-sophomore curriculum with more applied material. Examples of a refreshing new approach to the calculus, by way of applications, can be seen in the Amherst project materials [AMH] and the Harvard materials [HAL].

If you decide to work applications into your class then consider this. Mathematical modeling is complicated and difficult. If you take an already complex mathematical idea and spend an hour applying it to analyze a predator-prey problem, or to derive Kepler's third law, or to design the Wankel engine, then you are likely to lose all but the most capable students in the room. How will you test them on this material? Can you ask the students to do homework problems based on this presentation?

Think of what it is like to teach the divergence theorem. There are almost too many ideas, layered one atop the other, for a freshman or sophomore to handle. Students must simultaneously keep in mind the ideas of gradient, surface, surface integral, curl, and so forth. Most cannot do it. The same phenomenon occurs when one is attempting to get students to understand a really meaty application. I am not advising you against doing these applications. What I *am* saying is this: if you should choose to do one, go into it with your eyes open. If, after fifteen minutes, the students' eyes glaze over then you will have to shift gears. Be prepared with a physical experiment, or a film strip, or an overhead slide, or a transition into another topic.

An effective presentation of an application should be broken into segments: a little analysis, a little calculation, a little demonstration. Lower division students cannot follow a one hour analysis. Always bear in mind that, no matter how satisfying you find a particular application, your audience of freshmen may be somewhat less enthusiastic and may require help and encouragement.

On the other side of the coin, don't get sucked into doing just trivial, artificial applications. This cheapens our mission in the students' eyes and makes us seem less than genuine. The calculus and its applications are among the great achievements of western civilization. Be proud to share with the class the analytical power of calculus. Do so by presenting some profound applications, but put some effort into making the presentation palatable.

My experience is that, for freshmen, short applications are the best. They can be modern, they can be interesting, but if doing the analysis entails layering too many levels of ideas on top of each other then most students will be lost. And there is always the danger that some student will ask the question "will this be on the test?" What can you say? Will an hour-long application be on the test? No, but some of the analytical techniques that you use in the example could be. You had better have an answer prepared for questions such as these. See also Section 3.5 on answering difficult questions.

William Thurston, in his article [THU] on the teaching of mathematics, points out that mathematics is a "tall" subject and that mathematics is a "wide" subject. The tallness articulates the fact that mathematics builds up and up, each new topic taking advantage of previous ones. It is wide in the sense that it is a highly diverse and interactive melange. It interfaces with all of the other sciences, with engineering, and with psychology and many other disciplines. It is our job as teachers of mathematics to introduce students to this exciting field, and to motivate students to want to study mathematics and to major in it. Applications are a device for achieving this end. Using them wisely and well in the classroom is a non-trivial matter. Talk to experienced faculty in your department about what resources are available to help you present meaningful applications to your classes.

CHAPTER II

Practical Matters

1. Voice

There is nothing more stultifying than a lecture in a reasonably large class on a hot day delivered by an oblivious professor mumbling to himself at the front of the room. We are not all actors or comedians or even great public speakers. But we are teachers, and we must convey a body of material. We must capture the class's attention. We must *fill the room.*

I am not saying that you must lose your dignity, or act silly, or show off. You must learn to use your voice and your eyes and your body and your presence as a tool. If you are going to say something important, then make a meaningful pause beforehand. *Say* that it is important. Repeat the point. Write it down. Give an example. Repeat it again.

You can gain the attention of a large group by lowering your voice. Or by raising it. Or by pausing. One thing is certain: you will not gain the audience's attention by rolling along in an uninflected monotone. Again, I am not suggesting that you undergo a personality change in order to be an effective teacher.

At a well-known university in southern California they once tried bringing in actors from Hollywood to help professors spice up their delivery. Such pandering is inappropriate, offensive, and childish.

What I *am* suggesting here is that you take just a little time and contemplate your lecture style. A lecture or class should be a controlled conversation with your audience. It is a trifle one-sided, of course. But there must be cerebral interaction between the teacher and the students. That means that you, the instructor, must grab and maintain the attention of the class. Your behavior in front of the group is a primary tool for keeping the lines of communication open.

When you are talking about a subject that you perceive to be trivial, and when you are nervous, you tend to talk too fast. Novice instructors find themselves barreling through their lectures. You must resist this tendency. If you are really new at the business of teaching, then practice your lectures. Get a friend to listen. In calculus, a fifty minute lecture with four or five good examples and some intermediate explanatory material is probably just about right (I'm thinking here of a lecture on max-min problems, for example). Try to make each lecture consist of about that much material, and make it fill the hour. If you finish early, that is fine (but it may mean that you talked too fast). You can quit early for that day, or do an extra example, or use the extra time to answer questions.

Don't give the students the impression that you are in a rush. It puts them off, and reflects a bad attitude toward the teaching process. If on Wednesday you plan to explain the chain rule, then do just that. If the chosen topic does not fill the hour, then do an extra example or field questions. Do not race on to the next topic. One idea per lecture, at the lower division level, is about right. [Of course if you are teaching a multi-section class at a big university, then it is important to keep pace with the other instructors. This is yet another reason for keeping careful track of your use of time. See also Section 1.11.]

It is something of an oversimplification, but still true, that a portion of the teacher's role is as a cheer leader. You are, by example, trying to convince the students that this ostensibly difficult material is do-able. Part of the secret to success in this process is to have a controlled, relaxed voice, to appear to be at ease, and to be organized. Don't let a small error fluster you. Make it seem as though such a slip can happen to anyone, and that fixing it is akin to tying your shoelace or pulling up your socks.

But, as with all advice in this booklet, you must temper the thoughts in the last paragraph with a dose of realism. If you make the material look very easy, then students will infer that it *is* very easy. The psychological processes at play here are not completely straightforward. Nobody would be foolish enough to go to an Isaac Stern concert and come away with the impression that playing the violin is trivial. Yet students attend my calculus lectures, watch me solve problems, conclude that the material is easy and that they have it down cold, decide that in fact they *don't* need to do any homework problems or read the book, and then flunk the midterm.

These are the same students who come to me after the exam and say "I understand all the ideas. The material is absolutely clear when you talk about it in class. But I couldn't do the problems on the exam." I like to tease my students by reminding them that this is like saying "I really understand how to swim, but every time I get in the water I drown."

On the one hand, you don't want to make straightforward material look hard. After 300 years, we've got calculus sewn up. There is no topic in the course that is intrinsically difficult. We merely need to train our students to do it. So *do* make each technique look straightforward. But remind the students that *they themselves need to practice.* Do this by telling them so, by giving quizzes, by varying the examples and introducing little surprises. Ask the class questions to make the students turn the ideas over in their own minds. Use your *voice* to encourage, to wheedle, to cajole, to question, to stimulate.

Even if you know how to use your voice effectively with a small audience, there are special problems with the large audiences that occur in the teaching of calculus (for instance) at many universities. Refer to Section 2.13 for more on this matter.

2. Eye Contact

We all know certain people who invariably emerge as the leader of any group conversation. Such people seem to sparkle with wit, erudition, and presence. They have a sense of humor, and they are intelligent. What is their secret?

The answer is multi-faceted. This is obviously a talent that you must cultivate. Part of the trick is to show genuine interest in what other people have to say before bounding ahead with what you have to say. Another part is to talk about subjects, and to tell anecdotes, that you know will interest other people. Nothing is more boring, for instance, than listening to a half hour discourse on collecting firewood in the forest if you yourself are not disposed to this activity.

Many of the devices that make for an engaging conversationalist also make for an engaging lecturer. A review of the last paragraphs, and the rest of this booklet, will bear out this assertion. The device that I want to dwell on here is *eye contact.*

Telling a good joke while staring at the floor with your thumb in your ear will not have the same effect as telling the joke while looking at your listener, engaging his/her attention, and reacting to the listener while the listener is reacting to you. A good joke teller has his audience starting to chuckle half way through the joke and just dying for the punch line. Getting a good laugh is then a foregone conclusion.

Giving a good lecture is serious business, and is not the same as telling a joke. But many of the moves are the same. If you want to hold your audience's attention then you must look at your audience. You must engage not one individual but all. A good lecturer speaks to individuals in the audience, to grouplets in the audience, and to the whole audience. Like a movie camera, you must zoom in and zoom out to get the effects that you wish to achieve. A ninety minute movie filmed at the same constant focal length would be dreadfully boring. Ditto for a lecture.

Some people are very shy about establishing eye contact. It is a device that you must consciously cultivate. The end result is worth it: the lecturer who can establish eye contact is also the lecturer who is confident, who is well prepared, and who delivers a good lecture.

3. Blackboard Technique

Write neatly. Write either in very plain long hand or print. [Some people object to printing because of the clickety-click of the chalk. You'll have to make your own peace with this objection.] Be sure that your handwriting is large enough. Be sure that it is dark enough. Endeavor to write straight across the blackboard in a horizontal line. Proceed in a linear fashion; don't have a lot of insertions, arrows, and diagonally written asides.

Don't put too much material on each board. The ideas stand out more vividly if they are not hemmed in by a lot of adjacent material. In particular, it is difficult for students to pay attention when the teacher fills the board with long line after long line of neat print. An excellent guitarist once said that the silences in his music were at least as important as the notes. When you are laying material out on a blackboard, the same can be said of the blank spaces.

Label your equations so that you can refer to them verbally. Draw sketches neatly. Use horizontal lines to set off related bodies of material.

You can control your output more effectively by keeping the length of each line that you write short. Think of the blackboard as being divided into several boxes and write your lecture by putting one idea in each box. If necessary, *actually divide the blackboard into boxes.*

If the lecture hall has sliding blackboards, think ahead about how to use them so that the most (and most recent) material is visible at one time. For those combinatorial theorists among you, or those experts on the game of NIM, this should be fun.

If you are right-handed, consider starting at the right hand extreme of the blackboard space and working left. The reason? That way you are never standing in front of what you've written. Good teaching consists in large part of a lot of little details like this. You shouldn't be pathological about these details, but if you are aware that they are there then you will pick up on them.

Try to think ahead. Material that needs to be kept should be written on a blackboard to the far left or right where it is out of the way but can be referred to easily. You may wish to reserve a box on the blackboard for asides or remarks. This is another aspect of the precept that you know the material cold so that you can concentrate on your delivery. Just as an actor knows his lines cold so that he can make bold entrances and exits, and not trip over his feet, so you must be able to focus a significant portion of your brain on the *conveying of the information.*

If your lecture will involve one or more difficult figures then practice them on a sheet of paper in advance. Remember that you are a mathematical role model for the students. If you make it appear that it is difficult for *you* to draw a hyperboloid of one sheet, then how are the students supposed to be able to do it? Of course you can prepare the figure in advance on an overhead slide. This solves the problem of having a nice figure to show the students; it does not solve the problem of *showing* the students how to draw the figure. If necessary, consult a colleague who is artistically adept for tips on how to draw difficult figures.

If you cannot organize the steps of a maximum-minimum problem, then can you really expect the students to do so? In the best of all possible worlds, the students' work is but a pale shadow of your own. So your work should be the platonic ideal. Sometimes, in presenting an example or solving a problem, you may inadvertently gloss from one step to another; or you might make a straightforward presentation look like a bag of tricks. This is very confusing for students, especially the ones who lack confidence. By organizing the solution in a step-by-step format you can avoid these slips.

After you have filled a board, it should be neat enough and clear enough that you could snap a polaroid shot and read the lecture from the polaroid. In particular, you should not lecture by writing a few words, erasing those, and then writing some more words on top of the erased old words. Students cannot follow such a presentation. I cannot emphasize this point too strongly: Write from left to right and from top to bottom. *Do not erase.* When the first box is filled, proceed to the second. *Do not erase.* Only when all blackboards are full should you go back and begin erasing. Students must be given time to stare at what they've just seen as well as what is currently being written. Keep as much material as possible visible at all times.

Do not stand in front of what you are writing. Either stretch out your arm and write to the side or step aside frequently. Read what you are writing to the class. Make the mathematics happen before their eyes and *be sure that they can see everything. Say the words as you write them.* Every once in a while, pause and step aside to catch your breath and to let them catch up.

Here is a common error that is made even by the most seasoned professionals. Imagine that you do an example that begins with the sentences "Find the local maxima and minima of the function ... " And so forth. Say that you've worked the example. Now suppose that the next example begins with the same phrase. It is a dreadful mistake to erase all but that first phrase and begin the new example on the fly, as it were.

Why is this a mistake?—it *seems* perfectly logical. But the students are taking notes! How can they keep up if you pull a stunt like this? *Slow yourself down.* Write the words again. If a student gets two sentences behind then he/she may as well be two paragraphs behind. Give frequent respites for catch up.

And now a coda: how much of what you are saying should you write? In my experience, the answer is "As much as possible." When you are transmitting sophisticated technical ideas verbally, students have trouble keeping up. Many of them are not native English speakers. They need a little help. I write down everything except asides. I say the words as I write them. This is also a device for slowing myself down. Most of us tend to talk far too fast—at least about mathematics. Because of my poor handwriting, I *must* write deliberately when I lecture and this serves as an ideal counterpoint to my otherwise rapid speech. Each individual instructor will have to decide for himself how to strike a balance here.

There is a real psychological barrier for the instructor to overcome when learning blackboard technique, and voice control. When we understand very deeply what we are talking about, then it all seems quite trivial. We can convince ourselves rather easily—at least at a subconscious level—that it is embarrassing to stand in front of a group and enunciate whatever mundane material is the topic of the day. Thus we are inclined to race through it, both verbally and in the way that we render it on the blackboard. *Be conscious of this trap and do not fall into it.* I have never been criticized for being too clear, whether I was giving a calculus lecture or a colloquium lecture. *Slow down.* Be deliberate. Enunciate. Explain.

A mistake that we all make is to talk too much and write too little. Especially at the start of the hour, we tend to hyper-verbalize. Remember that this is technical material, the acoustics are not perfect, and many of the students are not native speakers. Writing will help you to slow yourself down and will help students to keep up. This is just a way of making yourself clear. Material that is rattled off verbally, and not written, appears to be unimportant. If it is important, write it down and *write it down clearly.* You need not write "Good morning class. Today is Wednesday." But write at least the key technical ideas that you are discussing.

4. Body Language

If you skulk into your classroom, stand slouching in front of the class with a furtive and disreputable expression, and are wearing slovenly clothing to boot, then you are sending numerous negative signals to your class. It sounds trite to say it, but dress neatly and attractively when you go to teach. Stand erect and look dignified. Attending to these mundane matters really does make a difference.

When you are teaching an elementary class, there is a tendency to suppose that everything you are saying is trivial. The upshot is that you talk too fast and write too fast *or not at all.* (See also Section 2.3 on Blackboard Technique.) Especially at the beginning of the hour, or at the start of a new topic, there is a great temptation to just rattle on (verbally) while writing little or nothing. This is a big mistake.

When you are teaching technical material to freshmen, it is impossible to be too clear. Write *everything* down. Write it neatly. Write it slowly. If a young student is six words behind then he/she may as well be six paragraphs behind. If you say "Recall that the interior extrema of a continuously differentiable function occur at points where the derivative is zero" then you have already lost most of them. Really. Write it down, at least in abbreviated form.

It is also the case that writing it down neatly and slowly is a subliminal way of telling the students that this material is important. If you are taking the trouble to write it down deliberately, then it must be worth writing deliberately. Conversely, if you scribble some incoherent gibberish, or scribble nothing at all, then what signal are you sending to the students?

5. Homework

In most lower division courses, and many upper division ones, it is by way of the homework that you have the greatest direct interaction with your students. When students waylay you after class or come to your office hour, it is usually to ask you about a homework problem. This is why the exercise sets in a textbook are often the most important part of the book (textbook authors do not seem to have caught on to this observation yet) *and* why it is critical that homework assignments be sensibly constructed.

Let me stress again that I am not trying to sell a time-consuming attitude or habit to you. If you take twenty minutes to compose a homework assignment then you are probably taking too much time. But consider the following precepts:

- Do not make the homework assignment too long.
- Do not make the homework assignment too short.
- Be sure that the assignment touches on all of the most important topics.
- Be sure that the homework assignment drills the students on the material that you want them to learn and the material that you will be testing them on.
- *Make sure that at least some of the homework problems are graded.*
- Plan ahead: the exams that you give should be based only on material that the students have seen in the homework.

If homework does not count and is not graded, then students will not do it. That is a fact. I realize that many of us have neither the time nor inclination to spend long hours each evening grading homework. Many universities and colleges these days simply do not have the resources to provide enough graders for lower division courses. But there are compromises that you can make. For example, you can tell the students that, of ten problems on the homework assignment, just three will be graded. But don't tell them which three. This device will force most of the serious students to do *all* the homework problems, but it requires much less grader time to get the grading done.

If the last suggestion will not work for you, then you can give weekly quizzes that you, yourself will grade. The amount of your time involved will be little, and it is a device to force students to keep up with the work. Incidentally, this device also gives you a gentle way to keep your finger on the pulse of the class.

Yet a third method is to have students exchange homework and grade each other's papers. I have never used this technique myself. It strikes me as having too many unpredictable factors in it, but it could work if used carefully.

Here is a problem/policy that you may want to consider. Students can and do benefit from collaboration, just as we mathematicians do in our research. While you probably do not want to encourage collaboration on exams, you may wish to encourage it on homework. Of course I'm not talking about "I'll copy yours this week and you can copy mine next week." Instead, I'm talking about an intelligent exchange of information among equals.

Some studies have shown that one reason that Oriental students in this country tend to do very well in their mathematics classes (and there are surely many reasons) is that they work in groups. More precisely, they first work hard as individuals; then they get together and compare results. In short, they collaborate much in the way that mature mathematicians collaborate. They are willing to say "I can do this but I cannot do that". Conversely, the studies indicate that certain other elements of the student population are either loath to work in groups or are unaware of the benefits of this activity; these groups tend to do poorly in mathematics classes. See [TRE] for details.

Some of the more interesting teaching reform projects, including those from Harvard and Duke, are specifically designed to encourage students to learn mathematics through group activities. Reports on these experiments are encouraging.

If you do decide to encourage group work in your classes, then you will have to make peace between said collaboration and your grading policies. If homework is not collected, then there is no problem and you can separate the good students from the bad through exams and quizzes. If instead homework is collected, then you will have to consider carefully how to tell whose work is whose, or at least how to divide up the credit.

6. Office Hours

At most universities the instructor is required to hold three or more office hours per week. Promise students that you will be there during that time. And be there. Students should be made to understand that they need not wait for a natural or personal disaster in order to come to your office hour. It is perfectly all right for a student to come to your office hour and say "I don't understand problem 6" or "the chain rule makes no sense to me".

You will usually not be overwhelmed with students (except perhaps just before an exam). In fact it is a general rule of thumb that, the larger the class, the smaller the percentage of students who will come to your office hour. But those who show up will appreciate your attentions. Of the hours that you have designated, you can spend some of them catching up on your correspondence, making up the next homework assignment, or reading the *Notices of the AMS* or the *Monthly*.

The office hour is your opportunity to get to know at least some of your students personally. This has several beneficial side effects, both for you and for them. When you are lecturing, you can have certain individuals in the room in mind as you formulate your remarks. You can make reference (*without* mentioning any names) to questions that came up in office hour. It is reassuring to the average student (the type that *does not* go to office hour) to know that good students (the type that *do* go to office hour) have some of the same questions that they have.

This point is in fact worth developing. Some components of teaching may be compared with certain components of psychotherapy. One big aspect of therapy—certainly an aspect that is exploited by popular psychology and self-help books—is to assure the patient that he/she is not alone. There are thousands of people with exactly the same problems, suffering in just the same ways. And they have been successfully treated.

Just so, when you teach you must give both subliminal and explicit reassurances to students that their questions and confusions are not theirs alone. An eighteen year old is scared to death that he/she is the only person in the room who doesn't understand why the numerator in the quotient rule has the form that it has—or why it does not seem to be symmetric in its arguments. Such a student would not dare ask about it in front of a room full of his/her peers; the student may not even be sure how to articulate the question, so surely will not want to flounder about in front of the entire class. At the same time the student may be afraid to come to your office hour and, alone but in *your* august presence, ask for a clarification.

Thus you must signal to students that questions are a good thing. When a student asks a question in class that might be of general interest, I not only repeat it but I often state that I am glad this question was raised. I carefully record the question on the blackboard. Several people have visited me privately, I add, and asked variants of the same question. If there is a question that should be asked but has not been, then I ask it myself. I say that if this point is unclear to them (the students) then they should come see me in my office hour and get it straightened out. You don't need to give away door prizes to drum up business at your office hour. However it is psychologically important for students to know that you are available, whether they actually come to see you or not.

I often announce to my classes that students may drop by my office even when it is not my office hour. If I am not busy, I'll be happy to talk to them. In practice this does not appreciably increase the flow of business. There are always students who strictly respect your designated office hour and there are always those who drop by when they please. But making an announcement of this nature is one of those little details that contribute to good teaching; for it sends a signal to the class that you care. If you do make such an announcement, be courteous to those who take you up on it. If you are busy and must send the student away, do so with respect and suggest another time for the student to return.

The office hour is a way to step out of your role as lecturer and let the students know that you are a person. It is a way to become acquainted with some of your students. Any good public speaker "works the audience" before his/her speech. Holding your office hour is one way to work the audience. You will also get a feeling during the office hour for how the class is doing, what problems and concerns have arisen, how the pace is working. It is wrong, and self-defeating, to view your office hour as a dreary duty; it is a teaching tool that you should use wisely.

7. Designing a Course

This topic is a part of the subject of "curriculum"—there are those in schools of education who earn advanced degrees in curriculum. It is a big world. I simply want to lay down a few simple precepts.

First, if the course you are teaching is a service course (i.e. one whose audience consists in large part of engineers, economists, or other non-math types), then you probably will *not* have a lot of leeway in designing your course. Certainly the content will be pre-specified. And, like it or not, you had better stick to the syllabus or outline that the department provides for you. Your students are taking this course *only because* it is required for their major. If you get a reputation around campus as the instructor who louses up this course for students (so that they must repeat it), then before you know it other departments will be designing their own course to substitute for this one. Such a consequence of your actions will not endear you to your department head. Funding and hiring at many universities is linked rather directly to the number of courses offered and number of students taught. You do not want to be known as the instructor who killed Math 117.

Even if the course you are teaching seems to be good old fashioned pure math, and is taught almost exclusively to mathematics majors, you may find that your department has rather rigid ideas about what should be included in the course. The department may also have a committee that selects the text. Don't be afraid to approach a more experienced faculty member for advice on this matter (see also Section 2.16).

If indeed you are teaching a course—perhaps an advanced topics course—where you will have a free hand, then the main precept is to *slow down*. It is too easy to fall into the trap of preparing the course for your fellow faculty (don't worry, they will almost never show up). This is another good opportunity for obtaining advice from a more experienced colleague. Draw up a tentative outline for the course and show it to a friend. Design the course in such a way that there are several natural places to quit for the semester. Don't drive students away with a syllabus that reads like "Everything that I know, or wish that I knew, about mathematics." Generally it is better to slow down, to give more examples, and to achieve more depth than to set up a situation that allows you to go to the next conference and tell of all the topics that you covered and all the others that you wanted to cover but could not.

Most students do not have the maturity or experience to look at a completely unreasonable course syllabus (see Section 2.11) and reason that "the instructor will never cover all this; of course he/she will slow down." You have to provide the leadership, both in the small and in the large.

Your reading list also sends important signals to your class. If your required reading is *Linear Operators* by Dunford and Schwartz and the complete works of Leonhard Euler, then (no matter what the merits of these books, or of your course), you will have no audience. Use a little practical sense when selecting a text. Also remember that, although most of your students love knowledge, they will not spend $250 for a book.

8. Handouts

It is tempting to write up a lot of handouts for your class. If you give a lecture on Stokes's theorem and feel that you have not made matters clear, then you might be inclined to draw up a handout to help students along. You also might suspect that this extra effort on your part will improve your teaching evaluations and, in particular, that students will appreciate all this additional work that you have put in. Well, it won't and they don't. Only prepare a handout when it will really make a difference. Students feel that they have enough to read already. Inundating them with handouts will only confuse them.

However let me temper this with one important exception. At many universities, it is common to distribute prepared lecture notes. At some, such as UCLA, the student association hires graduate students to prepare careful lecture notes of key courses (such as calculus) and sells them in the student store. This can be a real boon to the student. First, many students are unable to take good notes and listen to the lecture (and think!) at the same time. Knowing that good notes are available for a modest price gives such a student the freedom to sit back and really listen. Second, having prepared notes available makes missing class a less onerous inconvenience.

Having a lecture notes system is akin to providing students with a textbook. It does not really fly in the face of what I said in the first paragraph. You will have to use the advice from this section in the context of what resources are available at your institution.

9. Exams

In most elementary classes the principal device for determining grades is the examination. These are usually (but not always) given in class, or during a special time slot in the evening. There are a number of points of view about what constitutes a good exam.

Some professors attempt to put together elaborate problems, each of which synthesizes several of the concepts introduced in the course. This causes me to introduce a question which you should ask yourself frequently when you teach or write: "Who is my audience? Am I trying to teach eighteen year olds or am I trying to impress myself? Am I trying to effect an educational experience? Or am I trying to put together an exam that I can show to my cronies while crowing about how dumb it proves the students to be?"

By contrast, it is also possible to carry minimalism to an extreme. A famous exam from MIT consisted of the single problem: "You have a pile of warm metal shavings in the shape of a cone. Discuss." This may have been appropriate for students at MIT in the '50's. It is not appropriate at most universities today.

My attitude is extreme in yet a third direction. I usually tell my students what will be on the exam. No, I don't write each exam problem on the blackboard during a review session. But if a student asks "Will we be tested on the chain rule?", I give him/her an honest answer. If the student says "How many problems are on the exam," I tell. If a student wants to know how many questions are multiple choice and how many not, I give. To deny this information is just power tripping. It serves no good purpose.

Exam time is when you really have the students' attention. Get as much as you can from it. Drive home what the important ideas of the course are. Give a thumbnail sketch of the evolution of these ideas a few days before the exam. It helps students to organize their thoughts.

To be honest, 95% of my exam questions (in an elementary course) are straightforward. They offer no surprises. They are similar, but not isomorphic to, homework exercises. With the other 5% I am more fast and loose. I use these as a vehicle to identify the really bright and able students in the class.

I know good teachers at first class universities who take the straightforward approach one step further. They have a blanket policy in all elementary classes (calculus and linear algebra, let's say) that *all* exam questions come from the homework. Literally. And they announce this on the first day of class and repeatedly throughout the course. It's an interesting policy: they tell the students exactly what will be on the test (in a sense), but on the other hand they really don't. This policy leaves students little room for complaining about the content of exams. On the other hand, it does not challenge them.

My message here is that there is no sense in alienating the class, and there are so many ways in which you can do so. Many of these alienation devices are inadvertent, so why go out of your way to offend your students? And, again, if you are consciously going to give your students a killer exam then you should ask yourself *why* you are doing it. What are you trying to accomplish? Whom are you trying to impress? Consider carefully before you give such an exam. If the class is already dead then giving a hairy exam will pound the final nail into the coffin's lid. If the class is on your side, then why make a conscious effort to drive the students away?

Put another way, the purpose of a class is to transmit knowledge and information. Any given class has a dozen or more key ideas in it. That is what the tests should be about. A midterm or final exam in a basic course should not be a repository for ancillary theorems; it should not be a forum for obscure results not covered in class, or touched upon only in passing. An exam should be about the principal topics in the course. Topics covered on the exam should be ones that the students have heard about in lecture and seen in the homework.

Make sure that the questions you ask elicit the basic information that you seek. If your question about the chain rule turns into an algebraic morass, then it does not test the students about the basic material that they are to have mastered. If your maximum-minimum problem involves arithmetically or algebraically complex expressions that obscure what is going on, then you are not really testing the students as you wish to do. Thus it is important that you work the problems through in advance. This takes some time, but less time than all the aggravation that ensues if you give a poor exam.

Multiple choice or show the full solution? There are arguments for and against both systems. From the professor's point of view, one argument for multiple choice is that the grading of the exams requires no effort (in many cases it can be done by machine). And the exam is completely objective. But these reasons are a bit self-serving, and there is another more interesting consideration:

If you give traditional exams on which students write out solutions to the problems, then you usually fall into the malaise when grading of giving a lot of partial credit. Since you are human, you may tend to give even more partial credit on the 75th examination paper than on the 5th. The upshot is that it is actually possible for a student to get through the entire calculus sequence, with a grade of "C" or better, not knowing any particular calculus technique in its entirety. By contrast, it can be argued, the multiple choice exam has the advantage of requiring the student to actually *get to the correct answer* on a number of problems. But there is more to mathematics than just getting the correct answer; so you must consider to what extent your multiple choice exam is exposing students to the wrong value system.

On the other side, it can be argued that multiple choice exams involve a lot of gamesmanship. A student who has not studied, but who is clever, can sometimes get a reasonable grade on such an exam just by guessing shrewdly. [Of course you can offset this with negative scores for wrong answers. Also, if you give about ten possible choices for each question, and if the exam is otherwise well constructed, then you can make this eventuality unlikely.] It can also be argued that it is easier to cheat on a multiple choice exam.

Perhaps more critical these days is that multiple choice exams do not appear to be a good vehicle for training students to do multiple step problems. This is one aspect of mathematical training in which American students lag behind students in Japan and other countries.

I find it useful to compose my exams in elementary courses as follows (this is assuming that we are dealing with a really large class for which a traditional hand-graded exam is out of the question): If there are twelve problems on the exam then ten of them are multiple choice and two are "short answer." The short answer problems are of the sort that I can grade instantly by just glancing at them.

The students in large classes that I have taught are comfortable with an exam that is primarily multiple choice. But they enjoy the personal touch suggested by a couple of short-answer problems that are hand graded.

It seems to me that in a small class the professor can write a traditional exam requiring full answers to questions and then spend some time grading the papers carefully. In this context you can not only attend to the grading yourself but can make constructive comments. These comments can be brief, and they can be encouraging. The serious students do read them, and do benefit from them.

I have presented arguments in favor of machine-graded multiple choice exams and also arguments against them. Once again, I shall be prescriptive: hand-graded exams are better. They keep you in touch with how the class is doing as a whole, and also with individuals in the class. They give you the opportunity to discern what topics require additional coverage in class. Your comments on the exam are a useful part of the teaching process. If it is at all feasible, even in a class of eighty or more students, endeavor to give traditional hand-graded exam.

It is tempting, especially for new instructors, to hold review sessions for exams. This is a way of making yourself feel like Santa Claus, it easier than doing something more productive, and it makes the students grateful. But it also makes exams seem more onerous than they really are. [If you do decide to hold a review session anyway, then read Section 2.14 on problem sessions.] And it makes the students who cannot attend the review session feel as though they are at a serious disadvantage. I find it more effective to write a practice exam that I distribute a week in advance of the real test. About two days before the real test I post solutions of the practice problems. Of course there is always the danger that students will think that first reading the practice problems and then reading your solutions will constitute studying for the exam. I always caution the students strenuously against this trap. No system is perfect.

Tests that are too long, or too involved, do not work. Your exam should contain a reasonable number of questions of reasonable length, and they should not be interlinked. If problems are interconnected and a student makes a critical error in one of these then all of the related problems are affected. If test problems are too involved then students can panic, mismanage their time, and turn in a performance that does not at all reflect their true abilities.

It is very easy to misjudge a test that you write. A problem that seems trivial at first blush may have complex arithmetic or algebra hidden in it. Thus *you must personally work the test out completely before you give it to your class.* An exam that you can do in ten or fifteen minutes—with all solutions written out neatly—is probably about right for a 50 minute exam for a class of freshmen. If it takes you 40 minutes, and you find yourself laboring over the algebra or arithmetic, then obviously this is not a suitable 50 minute exam for freshmen.

The point value of each exam question should be clearly exhibited on the exam. The total number of possible points on the exam should be displayed. It is tempting to make difficult problems worth a lot of points and trivial problems worth very few. But of course the end result, since many students will not do well on the hard problems, is that the class average is pushed down. On the other hand, you don't want to make the easy problems worth a lot of points and the hard ones worth just a few—this sends entirely the wrong message to the class about what is important. So you must strike a balance.

It is a useful device to break difficult exam questions up into steps. This helps the weaker students to get started, and to display what portion of the material they actually know. It also makes the exam easier to grade, and increases the consistency of your grading.

When you are grading exams, it is important to be as consistent as you can be. Begin by writing out the solution to each problem. Break the solution into pieces and assign a point value to each part. Thus, in a maximum-minimum problem, setting it up might be worth 3 points, doing the calculations another 3, and enunciating the answer another 3. One spare point for overall analysis makes a total of ten. Some instructors like to be more precise than this. Refer to Section 2.13 for the concept of "horizontal" grading for insuring uniformity in grading.

Remember that some students will come to you with questions about how their exams were graded. In some cases, they will come with a friend and ask why two similar solutions were graded differently. If you are systematic, then you can handle such transactions with dispatch.

When teaching a big class, it is best to generate some statistics about each exam that you give. When you hand an exam back to 200 people and someone asks "What is the average?" or "What is the cutoff for an 'A'?" then you had better have an answer ready. The alternative is chaos.

Incidentally, hand exams back at the *end* of the class period; for if you return them at the beginning of the hour then students will spend the period reading the exam and comparing grades rather than listening to your lecture.

It seems natural to spend the class period following the exam actually working the exam at the board. Let me tell you decisively that this is not a good use of time. First, students resent the implicit statement you are making to the effect that "look at me—unlike you, I can do the exam quickly." Second, what each student really cares about is how he/she performed on the exam. If a student did a problem incorrectly, it is possible that he/she will want to see you give the correct solution. Otherwise interest is minimal.

Best is to write out solutions on paper and either (i) distribute a copy to each person in the class or (ii) (for a large class) post the solutions in an easily accessible location.

Generally speaking, it is best to try to deal with specific student questions about the way a specific person's exam was graded in a private, one-on-one fashion. You should never handle such complaints in front of a class. It is also a bad idea to handle them in front of a group of six after class. This is a personal matter. Treat it like an appointment with a physician. See also Section 2.10 on grading.

10. Grading

The pot of gold at the end of the rainbow, from the student's point of view, is the grade at the end of the course. Grading is a multi-parameter problem; there are a variety of devices for making your grading scheme more palatable (without being essentially more lenient) to students. What is the most even-handed and efficient way to determine grades?

I have used a number of grading devices successfully (and some unsuccessfully). I would like to record a few of the former here—merely for the reader's delectation. My main goal in formulating my grading policies is to make the greatest number of students feel that they have been treated fairly (and, not incidentally, to reduce student complaints). This does not mean that I am a lenient grader, nor that I give away grades for no special reason.

One device that I have used in large calculus lectures is the following: I tell students that I determine their grade by weighting their midterms as 50% of the grade, the final as 30% of the grade, and the homework as 20% of the grade (for example). But the *caveat* that I throw in is that anyone who gets an "A" on the final gets an "A" in the course. This assumes, of course, that the final exam is comprehensive. Thus if a student comes to you during the term and is distraught about his/her homework grade or his/her midterm grade, then you can simply enjoin that student to do well on the final. In fact not many students pull their grade up with the final exam (never more than 5%) and this simple device helps to keep morale high.

Always tell students on the first day of class, and in your syllabus (see Section 2.11), how you will grade the course. You want this to be a matter of public record. If students complain about your grading practices, and there are occasionally some who will, then you have your public statements to fall back on. And don't lie. If you say that you will grade by a certain method then do so. If you say in your syllabus that you will grade on a curve, then do so. If you say in the syllabus that you have an absolute grading method (90% is an "A", 80% is a "B", etc.) then stick to that.

Try to set up your course grading scale in such a way that, if a student comes to complain that he/she missed an "A" by only five points, then you can counter that the student has already benefited from certain built-in largesse. This largesse could include adding five points to everyone's total at the end of the course.

An alternative device which has always worked for me in borderline cases on, say, a midterm grade is to say "If these few points really make a difference in your course grade at the end of the semester then come see me then." This usually makes everyone happy, and very few students will take up this offer.

If a student comes to complain about a grade, then show the student some courtesy. If you cannot come up with a cogent reason for the way that you graded an exam or a problem, then that is *your fault*. Rethink your grading practices. Never fall back on your august position as your first line of defense. You show your students absolutely no respect to say "that's the way I graded your test and I'm the boss." That is not how you would wish to be treated. You can always turn a session to analyze how a test was graded into a favorable transaction. What does it cost you to give the student a few extra points if the points are merited?

However *never* penalize a student for being honest. If the student comes to you and says "you added up my points incorrectly: I should have received an 87 instead of a 90" then just send the student home with a little praise for being so perceptive. Tell the student that if you ever inadvertently give too few points then he/she should feel free to approach you at that time as well.

It is tempting, especially when you are a new instructor, to try to take an organic approach toward grading. Students are very receptive to the instructor who says "I try to grade on a subjective system. If your strong grades are on the midterm exams, then I down-play the homework and the final. If your work shows improvement, then I take that into account when I determine your grade. I try to emphasize everyone's strengths. I am your friend." This approach works well in the short run. It is a good opiate—for you as well as for the students.

However the down side is that if a student complains about his/her course grade then you have nothing to fall back on. You cannot show him/her your calculations and you cannot show where his/her score fits on a histogram; the trouble with an intuitive methodology is that you cannot explain it or defend it. Even though it sounds a trifle cold, you are much better off with an objective system of grading. In the end, everyone is more comfortable with a dispassionate approach. And, in the rare event that you have to defend yourself to the chairman, or to an angry parent, or to the dean, or even to a colleague, you will be prepared.

Sometimes you must change a student's grade—either on an exam or in a course. Perhaps a clerical error was made in recording the grade, or an error was made in grading a problem, or any number of other human frailties can enter into the picture. Do not be afraid to change a grade when it is merited. *However*: you do not want to develop the reputation among students as an instructor with whom grades can be negotiated. I've had this rep; I don't know how I got it. But there was a time when, the day after each exam, 85% of my students lined up outside my office to take turns slugging it out with me—point by point and problem by problem—over their exam grade. This is unpleasant and (can be) degrading both for you and for the student. Doing a careful job of grading in the first place, and posting carefully written solutions for students to see, can help to assuage much of student discomfort with grades.

Many times I, as an inexperienced instructor, have spent fifteen minutes haggling with a student over a problem only to realize that the student had not read the correct solution. *Make the student read your solution before you agree to talk about the grading of a problem.*

11. The Syllabus

Every mathematics course should have a syllabus—for the same reason that every textbook should have a preface. The teacher (author) should have given the course (book) a little thought and planning before the course (book) starts. What text will be used? How much material will be covered? What is the assumed background of the audience? How many exams will there be? How will the grade be determined? What will be the pace of the course? What is the instructor's name, office number, phone number, and office hour?

It is only courteous to provide students with this information. The syllabus is also a paper trail for your course. Try to stick to your syllabus as much as possible. If a student comes to you in the seventh week and says "I didn't know that there would be a midterm on this date" or "I didn't know that homework was such a big component of the course" then you can point out that this was explained in the syllabus that you distributed on the first day of class. The syllabus serves as a sort of contract between you and the class: it keeps you honest and it keeps the students honest.

The syllabus should not be a *magnum opus*. In some large calculus lectures, especially when there are several such lectures being coordinated, instructors may find it useful to list the topic for every lecture, relevant pages from the book, and the homework problems that are assigned for that day. That is fine in its place. For most courses a syllabus of one or two pages is more than sufficient.

Ideally, the syllabus should be available in a stack outside your door, or perhaps in the undergraduate office, all semester long. This is just good business, like a restaurant posting its menu or a gas station posting its prices. There is admittedly a certain cachet to being completely disorganized and doing everything by the seat of your pants, but it doesn't pay. You end up wasting a lot of time covering your tracks, you create too many potential opportunities for aggravation, and it leaves students with a bad taste in their mouths.

12. Choosing a Textbook

Choosing a textbook sounds like a simple motor skill. It is not. As a mathematician, you are trained to be rather self-indulgent. You may tend to choose a text that pleases you. If there is a cute new proof of Stokes's theorem or if there are four color computer-generated graphics, or if the book sales representative takes you to a particularly nice place for lunch, then you are liable to be influenced.

Remember that, while teaching the course, you will not spend nearly as much time reading the book as will the students. Try to step outside yourself as you look at a text. Ask yourself whether the students will be able to understand it, and whether they will be well motivated to try to do so. Are the examples well chosen, sufficient in number, and do they order from the simple to the complex (this device is known as "stepladdering")? Are the exercises suitable, and well coordinated with the text, lectures, and examples?

If you choose a bad text, you will pay for it throughout the semester. Homework assignments will be difficult to design. They will be difficult to grade as well. Students will not perform well on exams if they do not have a good book to study. If the text is weak then the correspondence between it and your lectures will be tenuous. This dilemma defeats class morale.

Here is an example of a problem that exists in the current crop of calculus texts in this country: What does mean to say that a function is discontinuous at a point $x = a$? Some calculus books state that a function is discontinuous at any point at which it is undefined. For example, the function $f(x) = 1/x$ is discontinuous at $x = 0$. But wait, it gets worse. Consider the function $g(x) = (x^2 - 1)/(x - 1)$. On the face of it, this function is undefined at $x = 1$ and hence discontinuous there. After division, however, the function becomes $x + 1$ which is both defined and continuous at $x = 1$. But are $(x^2 - 1)/(x - 1)$ and $x + 1$ the same function? Who knows? And beware: if you declare a convention different from that in the text, many students will become confused and some will be lost.

This is but one instance of the sort of mess that a textbook can get you into. Some people never use the first edition of a text because they reason that (i) all misprints will not be weeded out until the third edition and (ii) glitches like that described in the preceding paragraph will be mollified (in response to user protest) in later editions. In point of fact, the conundrum described in the last paragraph occurs in the third edition of a very successful calculus text. The (very famous and wealthy) author insists that this is the only way to handle continuity at a point where the function is undefined.

Let me expand on this last point. If a textbook uses notation or other conventions that you do not like, then don't use that book. You really are obliged to follow the notation in the book; otherwise all but the gifted students will be lost.

Now you can hardly be expected to read and digest every word of a text—especially one as cumbersome as a calculus book—before you use it in a class. The safest policy when selecting a text is to consult someone who has already used it. As with all teaching dilemmas, don't be afraid to seek the advice of a more experienced faculty member when selecting a text—especially if it is for a class that you have never taught before.

And now a coda on cost. It is not impossible these days for a textbook to cost $100 or more. If the book that you choose is expensive, be prepared to defend your choice. Is there an equally good book that costs just half as much? (I'm not talking here about *your monograph* on *your subject*; rather, I'm considering something like a linear algebra text, for which you should have less emotional involvement.) Checking the cost of the text for your class is just the sort of courtesy that you would have expected when you were a student.

13. Large Lectures

At many large state universities, and some private ones, it is common these days to teach some or all lower division courses in large lectures. Such a classroom situation offers special teaching problems. How can you, as the instructor, make yourself heard? How do you fill the room? How do you field questions? Do you really want the students to ask questions? If so, how can you encourage the students to do so? What about exams? How do you organize your teaching assistants?

Most large lecture halls come equipped with a microphone for the lecturer. Unless you have a voice like William Jennings Bryan, use it. If you don't use the mike, then either you will not be heard or you will be under such a visible strain that it will detract from what you are trying to accomplish. You *can* learn to become relaxed with a microphone clipped to your collar. When you have done so, you will be able to speak in a normal voice and be heard clearly. You can then concentrate on your mathematics.

I find that one way to develop student involvement in the classroom is to encourage questions. However, this device must be managed wisely in a large class. Too many questions will bring the lecture to a grinding halt. But a good one can make everyone prick up their ears. How can you get students to ask questions? Well, you can bait them. You can begin by saying "Are there any questions? Do you all understand the example that I just presented?" If that doesn't work you can say "Are you sure? Do you mean that if I gave you a pop quiz right now you would all get an 'A'?" If you do this with the right playful attitude, you will at first elicit an embarrassed giggle and then some hands will go up. Once you have created an atmosphere in your class in which people feel comfortable asking questions then this matter takes care of itself.

But here is a trap: Left to their own devices, students will lapse into asking questions of the rote form "How do you do problem 6?" Such questions must be discouraged in a large class, as they are boring and non-instructive. What you want are questions like "Why don't we define 'continuity' this way?" Or "Why does the chain rule have this form rather than that form?" It is up to you to *prompt* the students for the questions that you want. In order that they not become rhetorical questions, you must put these issues to the class and then *wait for an answer.* It is not enough to say "'Why does the product rule have this form? Well, here's why." If you want a reaction from your class, you must draw it forth.

When a student asks a question, *repeat it* so that you can be sure that the entire class has heard it. Writing it on the board is often a good idea as well. This is sound policy even in a small class. If you do not repeat the question, then the interchange between you and the student becomes a private conversation. The rest of the class is excluded. Other private conversations will start up. You will have lost control.

But there is another important consideration to repeating the student's question: if the question is not optimally formulated (or just plain wrong) then the repetition gives you an opportunity to clean it up or rephrase it. Then treat the issue raised with respect and answer it directly.

It is especially important in the large lecture situation that students be sure that you will not belittle them or make them look foolish in front of the other students. Be prepared to make even a dumb question look smart. If the first question gives rise to a second then say something like "Let's go back to the lecture for a bit. I think it will clarify this point for you." Or you can say "This question session is getting a bit too specialized. Why don't you see me after class?"

One device that works very well if you can manage it is to make yourself available in the front of the classroom for fifteen minutes after class. Students feel much more comfortable talking to you when surrounded by their peers, and while their questions are fresh in their minds.

Never, ever get involved in a personal discussion of grades in front of a class of any size. If student T wants to know why problem n was given only p points, tell that student to see you after class.

It is imperative that the instructor for a large lecture course be extra well prepared. If you begin to get lost when doing an example or side-tracked with an incorrect explanation then you will quickly lose a large segment of the audience, a lot of talking will start to take place, and the room will soon be bedlam (see also Section 3.6 on discipline). Everyone has off days and makes mistakes, but you must strive in a large lecture to have the material down cold.

The teaching of a large lecture course offers complications of a special nature. Discipline and commanding attention are two of these that are treated elsewhere in this booklet. But there are others. You *must* have a syllabus for such a course. You *must* prepare your homework assignments and exams carefully—and well in advance. There are few things more unpleasant than facing down a hostile audience of 350 hungry freshmen right before lunch. Therefore do not give exams on which the problems don't work out; do not give homework assignments on which the problems don't work out; *do* plan ahead for *all* activities. Have a fair and objective system of grading. If a student comes to you with questions about grades, have a fair, consistent, and clear set of data to show the student.

If you are the professor in charge of a large lecture, then you will probably be in charge of a group of 2 to 10 or more Teaching Assistants (T.A.s)—refer to Section 2.18. You must exercise organizational skills with them as well. Meet with them once a week to be sure that they are covering the right material, are aware of upcoming assignments and exams, and to apprise them of any difficulties that have arisen.

If the T.A.s are helping you to grade an exam, then you must tell them how you want the papers graded. Grade exams *horizontally* rather than *vertically*. This means that you should put T.A. #1 in charge of problem #1 (on all exams), T.A. #2 in charge of problem #2, and so forth. This is the only way to insure some consistency. If you let each T.A. grade a stack of exams from start to finish, then you will have wild inconsistencies and many student complaints.

Horizontal grading is also a useful device even when you are grading a stack of just twenty exams all by yourself. It will discipline you, and will tend to make the job go more quickly.

14. Problem Sessions, Review Sessions, and Help Sessions

At many big universities, the large thrice weekly lectures in a lower division math course are supplemented by once- or twice- weekly "problem sessions" or "help sessions." Usually the lectures are delivered by a professor or instructor while the help sessions are staffed by graduate student teaching assistants.

Imagine that you are the graduate student in charge of a problem session. It is easy to fall into the trap of not taking the work very seriously. After all, student attendance at these sessions is poor in general and spotty at best. Students seem to be inattentive and their questions are often puerile. But the quality of any class or help session is largely influenced by the attitudes and efforts of the instructor. If your attitude is to treat the help session casually or carelessly then you will get correspondingly disappointing results. Consider giving weekly quizzes, sending students to the board, and other devices for livening up your problem session. I wish to concentrate here on more mundane matters.

It is arguable that it is more difficult to conduct a good problem session than to give a good lecture. For the problem session presents all the difficulties of a lecture, and more. At least in a lecture you are in complete control of the order of topics and can, if you wish, present them from prepared notes. In a problem session, if you really let the students ask what they wish, then you must be ready for anything. And you must be able to think quickly, on your feet, of the best way to present any given topic, give a hint on any problem, or handle any point of confusion. In a lecture you can always pull rank and say "there is no time for questions now; see me in my office hour" (I don't recommend that you say this very often, but it is an option that is available). But help sessions are for questions.

If you are a novice, then it is probably safest to view the help session in the most naive way: your role is to help students do their homework assignment for *that week*. Thus your preparation for a help session might consist of working all the homework problems for the week, or at least staring at them long enough to be sure that you know how to do them.

Be certain that the techniques that you present are consistent with those used in class and in the book. Some professors require their T.A.s to attend their lectures, just to insure this consistency. For that matter, some professors attend the problem sessions at least once per week *accompanied by the grader!* This requires some extra effort on everybody's part, but it shows real consideration for the student who has questions about the way that his/her homework was graded. It goes without saying that in order to use this technique effectively the professor will have to be well-coordinated with the grader on how he/she wants each homework assignment graded.

When you are helping with a homework problem that is to be handed in, don't give away the store. One reasonable answer to the dreary question "How do you do number 14?" might be "I'll do number 16 for you, which is similar." Another reasonable answer might be "I'll get you started; you do the rest." A third is "'Here is an outline of the basic steps."

The advice to the T.A. to work all the homework problems the night before is one that I tender hesitantly. I never do this, but I've been teaching math for twenty years. I am rarely surprised by any question in a calculus class or help session, and even if I am I can usually slug my way through whatever new features are present. If I am at a review session for an exam and a student presents a really difficult question then I always have the option of saying "That's an interesting question, but one that could never be put on the test. Let's discuss it privately."

In your first few years of teaching you will have to strike a balance between being thoroughly prepared (by working all problems in advance) and spending too much time on preparation (see also Section 1.4). Just remember that a large part of your job is (i) to show the students how to do the problems and (ii) to convince the students that the problems are do-able (by ordinary mortals). If you fumble around and act baffled by the problems then you are presenting a poor role model and, more to the point, doing your job badly. Students find appealing the fact that I can do all the problems and that, moreover, I invariably know where the difficult spots are and help to chart the way through them. This ability can only come with experience; it is what you should strive for.

15. Transparencies

I have already touched upon the topic of overhead transparencies. With the use of transparencies, you can cover more material. By using several overhead projectors, you can create an ambience similar to that achieved by several blackboards. By using color, overlays, photos from books, computer printouts, histograms, and the like, you can put on a dazzling display of information.

Of course this type of teaching environment requires a lot of preparation. It may take several hours to amass the information needed to present a one hour multi-media event like that described in the first paragraph. It is not clear to me that the chain rule will thereby be any better conveyed than with a piece of chalk in the hand of a skilled lecturer. In fact, I have some strong opinions about this matter that I would like to take this opportunity to share.

I want my students to take my lecture as an inspiration to go home, pick up a pencil, and do some math. If they get the impression, even subliminally, that doing math requires a bunch of high tech equipment and software, then they may be disinclined to do so. If instead they just see a lone instructor with a piece of chalk doing math, then they may conclude that they can do it too.

On the other hand, there are some pictures (try drawing a hyperboloid of one sheet without practicing) that are difficult to do on the fly. A prepared picture on a transparency can be a great help. With MATHEMATICA, you can not only render a beautiful picture of a three dimensional graph, but you may also exhibit it from several different perspectives.

It is just about impossible to illustrate Newton's method adequately with freehand sketches using a piece of chalk. A computer printout on a transparency, or even an animation using a PC, can be a great help. Likewise Simpson's rule, the Runge-Kutta method, and other numerical techniques can best be illustrated with a machine.

BUT: If you are trying to teach your students to graph—to assimilate information given by the first derivative, the second derivative, intercepts, symmetry, and so forth—should you use MATHEMATICA or similar software as a teaching tool? Should you teach them to use hardware to generate printouts and transparencies? I think not.

We do not want to teach our students to push buttons. We want them to think analytically. It has been argued that MATHEMATICA and similar software can be used to help students interact dynamically with the graphics: vary the value of a in the equation

$$y = ax^2 + bx + c$$

and watch how the graph changes. That is not what I want my students to learn. I want them to understand that, for large values of x, the coefficient a is the most important of the three coefficients. And changing its value affects the first and second derivatives in a certain way. And, in turn, this affects the qualitative behavior of the graph in a predictable fashion. *After* these precepts are mastered, the student may have some fun verifying them with computer graphics. But not before.

Graphing is one of the basic techniques of analytical thinking. The picture is not an end in itself. It is the understanding that comes with the creation and analysis of the graph that is our goal as educators. Being able to push some buttons and render a beautiful picture or transparency of a graph in $\mathbb{R}^3$ is *not* the same as understanding the information contained in that graph.

Likewise, showing the students a prepared graph of a surface does not teach them the analytical tools required to generate such a graph, nor does it teach the ideas needed to *read* such a graph. Thus my recommendation is to use prepared transparencies with restraint.

16. Mentors and Neophytes

I've had many mathematical mentors, but never a teaching mentor. I've learned how to teach by making virtually every mistake that can possibly be made and then trying to learn from those mistakes.

Having a mentor—someone experienced in the craft of teaching—can save you a lot of travail. Your mentor will not always give you the answers that you want, nor possibly any answer at all. But you can benefit from the mentor's experience.

A mentor can show you the ropes when you are writing a syllabus or exam, help you to choose the right text, explain when you can skip a topic in a course, suggest how to handle disruptive students, or tell you what to do if you are running out of time in a course. A mentor can help you curve the results of a test. To ask for help is nothing to be ashamed of.

Many mathematics departments now assign a teaching mentor to each new young faculty member. If yours is one of these, take advantage of your mentor's experience and perspective. You don't have to follow all of the advice that is tendered, but at least you will have grist for your mill. If you are not assigned a mentor, then take the initiative and approach a senior faculty member. He/she will probably be quite pleased and flattered to be asked, and you will have initiated a useful relationship.

One interesting proposal is to have, as mentors to math instructors, professors from the history department, or the English department, or some other department that is fairly distant from mathematics. Such a mentor will not be affected by math department politics, or by past injustices or disappointments meted out by the mathematics department, and should (in principle) serve as a breath of fresh air in getting a new instructor off on the right foot. You will probably not be able to find a mentor from a department in the humanities without some help from your department. If you wish to pursue this course, then take the initiative and seek that help.

17. Tutors

A commonly asked piece of advice, usually from a student having trouble in your course, is "should I get a tutor?" I have a very simple answer to this question: "No". It is almost unavoidable that a student will treat a tutor as a crutch. The student figures (at least subliminally) that by paying $20 per hour (or whatever is the going rate) he/she is *buying* knowledge. And now looms the specter that to my mind should be the benchmark for all educational issues: all learning of significant knowledge requires considerable effort on the part of the learner. This fact has not changed since Euclid told Ptolemy (over 2000 years ago) that "There is no royal road to geometry." Instead of just slugging a new idea out for himself, the student finds himself thinking "I don't get this; I'll have to ask the tutor."

I could go on about this at length, but I will try to restrain myself. Go to any athletic facility and you will see young people spending hours perfecting their free throw or their skate board technique or their butterfly stroke. They don't hire tutors for that; they also don't hire tutors for learning to play the guitar or learning to modify their cars. The reason is very simple. There is plenty of peer support for these activities; young people are highly motivated to be proficient at them.

Not so with mathematics, or with most other areas of erudite learning. A good math student must be self-motivated. In most instances, the hiring of a tutor is an attempt by the student to buy his/her way out of some work. I've been a tutor; it's a great way to make money. But in most instances it is not an effective learning device. You might find it helpful to refer to Section 1.13 on math anxiety in connection with these issues.

Of course there are exceptions to what I am saying here. Some students are too timid, or too slow, or too far behind to catch up without help. In such cases a tutor may be necessary.

It is a sobering thought to realize how different the students' point of view is from our own. There is at least one high quality large state university today where students routinely hire a tutor for each class that they take—*before they have even set foot in the classroom.* Clearly these students have convinced themselves that classroom instruction is inadequate, or that their own abilities are substandard, or they do not know how to study and require a surfeit of hand holding. On days when you think that teaching is a straightforward process, stop and ponder this matter.

In any event, do not hire *yourself* out as a tutor for a student in a class that *you* are teaching. It is inappropriate, it is tawdry, it is a conflict of interest, and it might get you into trouble with your department. The safest policy is not to tutor students in your college or university at all. The point is this: you are already being paid a salary by your institution to educate the students at that institution. To further accept tutoring money from the students constitutes double dipping.

Even having to recommend tutors can put you in a position of conflict of interest. Most math departments maintain a list of qualified people who can tutor for math courses. This is done as a service for the students, but it is also done as a service for the faculty. When a student asks you about tutors, send that student to the departmental office and the official list. It really is the best policy.

18. For Those Who Have Been a T.A., and Those Who Have Not

Being a Teaching Assistant (T.A.) provides some experience in being a teacher. But it does not provide much, and the background that it provides can be misleading.

When you are a graduate T. A. at a big state university, you are probably not your own boss. In most cases you work, alongside several other T.A.s, for some professor who is delivering lectures to a large audience. On alternate days, the class will be broken up into smaller "quiz sessions" or "problem sessions", and you will be asked to teach one or more of these. You will also be asked to help with grading, with other assigned activities, and (primarily) you will be asked to do what you are told.

Being told what to do lifts a great deal of responsibility from your shoulders. But this also means that a T.A. has never really taught. You've had some experience standing in front of a group, organizing your thoughts, answering questions, developing blackboard technique, and so forth. But you will have never made up an exam, written a syllabus, designed a course, given a course grade, or any of the dozens of other activities that figure significantly in the teaching process.

Conversely, if you have never been a Teaching Assistant (either because in graduate school you were on a fellowship that had no formal duties attached to it, or perhaps because you were educated in another country), do not despair. At least you are entering this profession with possibly fewer prejudices than are held by those who have stood as a T.A. before a hostile audience in this country. Perhaps reading this booklet will provide you with better information and a better outlook than having served as a T.A. under a professor who doesn't know how to or even care about good teaching.

Let me put an ameliatory note here: some professors are well aware of the down side of being a T.A. and attempt to compensate for it. They give their T.A.s more responsibility. For instance, such a professor might write the first midterm exam for a class himself and then let the T.A.s write subsequent midterms (under close supervision). This is positive psychological reinforcement for the Teaching Assistants, and good experience for them as well. Likewise, the T.A.s can be allowed to set the curve for grading (under supervision) and to perform the other ordinary functions of the instructor. The professor is not being lazy here; rather, he/she probably has to expend more effort than if these tasks were done solo. But it provides awfully good experience for the graduate student T.A.

At some schools, the T.A. is more autonomous. It is possible that the T.A. will be a free-standing teacher, creating his/her own exams and constructing his/her own grading system. If this description applies to you, then this section does not. But the rest of the booklet does, and you may benefit from reading it.

For more information about the day-to-day duties of being a Teaching Assistant, see Section 2.14.

19. How to Ask, How to Answer

If a pollster asked the average American voter "What do you think of the upcoming election?", the resulting answer would probably not be very enlightening. If you turn to your calculus class one day and say "OK, now we've covered Chapters 5 and 6—any questions?" then you will get a bunch of blank looks. By the same token, if a textbook salesman hands a new calculus book to a math professor and says "What do you think?", the professor will probably say "I dunno; they all look the same to me." Conversely, students come to professors with questions such as "Like, you know; I don't think I understand any of this stuff we're doing."

It is a strange facet of the human condition that most of us don't know consciously what we think about most things most of the time. A skilled questioner learns to ask *specific questions* in order to get meaningful answers. Instead of asking your class if there are questions about Chapters 5 and 6, ask them if they are comfortable with the chain rule, or if they can do related rates problems, or falling body problems. The material in a person's memory is hung on hooks; you must reach for those hooks to get useful answers to your questions.

We implement this dictum naturally when writing an exam. You would never set an exam question for freshmen that said "tell everything you know about differential calculus." Instead you ask very specific questions. You want to train yourself to do the same when talking or lecturing to students. More, you want to train yourself to do the same in reverse when you are trying to elicit questions from students.

There is a gentle art of getting your students to pose questions. And I don't mean questions like "Will this be on the test?" I mean the kind of meaty, well-thought-out questions that we all live for. Perhaps the most common complaint that I hear from disillusioned mathematics instructors is that they cannot elicit participation from their lower division classes.

This is a non-trivial issue, and one to which an entire separate book could be devoted. Rather than tell you in detail how to elicit questions from your class, I am simply going to enjoin you to think about it. Consider wheedling, cajoling, joking, challenging, priming. Here is one clever technique that was devised by Jean Pedersen. She asserts that it works extraordinarily well for her: it is called the method of "mathematical POST-IT notes."

We all know that POST-IT notes are those little squares of colored paper that can be easily affixed or un-affixed to a document for the purposes of making remarks or memos. The idea for the application of these devices in a math class is that the professor comes to class with a tablet or two of these notes, each having the professor's name (or some other identifiable epithet) stamped on it. Whenever a student asks a good question, then he/she is rewarded with a POST-IT note. 'So what?' you ask.

When the next exam comes around, the students are instructed to bring their POST-IT notes along. They are to affix them to the front of the exam that they hand in. The student then receives two extra points (or some number to be pre-determined) for each POST-IT note.

Reports are that when this policy is announced in class, it is as though a jolt of electricity has run through the room. Suddenly hands are waving in the air, and previously uninterested students become the life of the party.

Now let me be the first to admit that this teaching device, like any other, is not perfect. Some students who are already alienated will become more alienated if they are unable to garner any POST-IT notes. Other students may object that they are being treated like children.

You should develop an arsenal of techniques for encouraging student participation in your classes. Here is another example that has been used successfully by one professor: On the first day of class this distinguished gentleman announces that he simply cannot spell. Students should feel free to correct his dreadful spelling. Then he begins to lecture, spelling "line" as "lien" and "book" as "buk". Students are so delighted to be able to confidently correct the professor's spelling that participating constructively in the mathematics portion of the course becomes very natural.

I find this last technique a bit dishonest, but it's hard to argue with success. I encourage you to experiment with different methods for bringing your class to life. I'm lucky in that there apparently is something about my demeanor in class that makes students comfortable with asking questions and participating. I believe that this fortunate attribute has evolved unconsciously. Others may need to give the matter serious conscious consideration.

As you experiment with ways to liven up your class, bear in mind the nature of the enemy. One enemy is that young adults, for the most part, are quite unsure of themselves. Unlike an experienced mathematician, who in effect makes a career out of asking (often stupid) questions, the student is deathly afraid of looking silly in front of his/her peers. Try to create an atmosphere in which you and the students are co-explorers; convey that you will sometimes make false starts, and so can they. It's a knack, but you can learn it.

Another enemy is that math *can be* (it is not by nature) a dry, forbidding subject. Part of your job as teacher is to make the students want to learn the material. This book supplies a variety of techniques for achieving that goal.

20. Incompletes and Other Gimcracks

The profession of teaching, while certainly a stimulating and rewarding one, is littered with nasty little details. One of these is the "incomplete." The theoretical purpose of an incomplete is to provide a vehicle for handling the following situation: a student *has* completed a substantial amount of material in the course, but has suffered a death in the family or some other setback; he/she needs to defer completion of the course work until the following term. Many universities find it convenient to let professors administer incompletes as they see fit; as a result, there is much inconsistency and abuse.

Frankly, I've given a lot of incompletes in my life and very few of them were ever completed. Students get busy with the next semester's work, and never get around to things past. In fact I did not complete the only incomplete that I ever took as a student. It is also unfortunately the case that certain students will just blow off a course and then ask for an incomplete at the end of the term. Often it is easier for you as the instructor to just grant the incomplete, given that an otherwise undisciplined student is not likely to complete it (the grade then usually, but not always, reverts to an "F"). You may very well wonder what is the point of engaging in a long interview with such a student to see whether the incomplete is merited.

All this having been said, it is probably best, as with all matters in teaching that impinge on fairness, to have a uniform policy for handling incompletes. But think this through: are you going to require that the student provide *proof* of his/her excuse? This sounds reasonable, but what if the student says "my mother is dying of cancer" or "my grandmother just died and I cannot concentrate on my work". I know professors who will demand a letter from the undertaker, but this strikes me as a bit extreme.

One convenient way to handle the request for an incomplete is to instruct the student to approach a professor teaching the same course the following term. The student should ask whether he/she can audit the course, having his/her work graded. Then the new professor will transmit the resulting grade to you. This is clean and simple, and it works. You certainly don't want to have to reteach some or all of the course for the benefit of just one student.

You are the academic analogue of a middle management executive in the business world. Executives exist, presumably, because they are smart enough to handle exceptional circumstances. Teaching is loaded with all of the sorts of exceptions that are connected with dealing with *people*. I have used the "incomplete" here as but one example of the problems and potential enigmas that can arise.

CHAPTER III

Sticky Wickets

1. Non-Native English Speakers

Many new instructors in this country are not native speakers of the English language. Besides all the usual stumbling blocks that are put before a new teacher, such a person has to worry about (i) lack of fluency with the language and (ii) lack of familiarity with classroom techniques in the United States.

If you are not a native speaker of English then fluency is entirely your responsibility. Every university offers a short crash course on English as a second language. Take the course. Watch television. Read books in English. Converse with American colleagues. If you are going to succeed as a teacher in this country, then you must speak the language well enough to (a) be understood and (b) be able to field questions. It is point (b) that causes more trouble than (a). Most students can get used to a lecturer who has less than perfect proficiency with the mother tongue. But if you as instructor cannot understand their questions then you will be a complete failure in and out of the classroom.

Some students are prone to complain, and will use any excuse to justify a complaint. I speak and articulate accentless English. Yet when I taught at Penn State some students complained that I did not speak with the local accent. Some students complain about teachers with British accents and teachers with Australian accents. There is nothing that you can do about such complaints so you should not worry about them.

What's true is this: if you are organized, if you speak up, if you treat students with the respect that *you* would desire from an instructor, and if you show some enthusiasm for what you are doing, then students will forgive a lot. Your foreign accent will fade into the background. Nobody will hear it any longer. And you will be a successful teacher.

Many professors who received their training in other countries are (justifiably) impatient with our students. American students do not specialize as early in their education as do, say, European students. Even in a sophomore differential equations class in America there is a broad cross-section of students that includes pre-medical students and others from outside the mainstream of mathematical science.

My message is this: learn to be patient. Students will ask you to repeat terms. Students might seem less able than those in your country. But they are bright and they are willing. You must learn to work with them. *After* you have learned how the American education system works, and what the students are like, you will find that your colleagues are receptive to your thoughts about its shortcomings. Before you have made this acquaintance, you are working in a vacuum and you should keep your own counsel.

In some countries it is the style of the university professor to stand at a lectern in the front of the room and to read the textbook to the class. Questions are considered a rude Americanism. An extreme example of a teaching style that is virtually orthogonal to what we Americans know is one that has been attributed to the celebrated Hungarian analyst F. Riesz. He would come to class accompanied by an Assistant Professor and an Associate Professor. The Associate Professor would read Riesz's famous text aloud to the class. The Assistant Professor would write the words on the blackboard. Riesz would stand front and center with his hands clasped behind his back and nod sagely.

My point is that in the United States, for better or for worse, things are different. The style here is to indulge in discourse with the class. Some professors make the discourse largely unilateral; that is, they lecture. Other professors encourage more interchange between the students and the teacher. Reading this booklet will help you to become acquainted with the traditional methods of teaching in this country.

2. Late Work

Late work is a nagging problem. The easiest solution to the "can I hand just this one assignment in late?" dilemma is to "Just say no." But what of the student who has a *really good excuse*? What if there has been a death in the family or some other crisis that the student cannot avoid?

The trouble with making one exception is that it tends to snowball at an exponential rate to N exceptions. In a large class this can be catastrophic. One possible solution is to tell the students that when you calculate their homework grade, you will drop their two worst grades. That means that any student can miss one or two homework assignments with essentially no penalty. It's a remarkably simple solution to an otherwise difficult problem.

There are a number of other possible answers to the late homework problem. You can downgrade late assignments, or you can assign extra work. You can just forget the missing assignment and base the student's course grade on the remaining course work. The point is that you should think about this matter in advance. A choice of incorrect policy toward late work could lead to a lot of extra work and/or aggravation for you. Ask a more experienced colleague for help in this matter.

3. Cheating

Cheating is a big, and probably unsolvable, problem. It is demoralizing for the teacher and for the non-cheating students. It is difficult to deal with the sort of students who cheat, for they may be dishonest with themselves and others in a number of aspects of their lives. You want to be firm and fair and just all at the same time.

You may wish to set a moral tone against cheating by making an announcement on the first day of class. For large lectures, this may be especially important and effective. Declare that you consider cheating to be an egregious offense, both against yourself and against the other members of the class. While you admit to the class that you may not be able to catch all cheaters, you assure the students that anyone caught cheating will be punished to the full extent of the law—*including expulsion from the university when appropriate.*

Be forewarned: most universities have set policies about handling cheaters. You are not free to act as you please when you catch a cheater. In particular, there are due process procedures set up (to protect the rights of the accused cheater) that you must follow if you wish to punish a cheater effectively. You do not necessarily have the right to tear up the student's exam, to give the student an F, or to mete out other retribution. Check with the director of undergraduate studies in your department to determine the proper course of action when handling a suspected cheater.

One rule of thumb is this: do not be lenient with cheaters. Cheating cuts at the very fiber of what university education is about. When you catch a cheater, you must send a strong message that this behavior is intolerable.

The best defense against cheaters is offense. Give your exams in a large room. Space the students far apart. Check picture ID's to make sure that students have not sent in ringers (substitutes) to take the exam for them. Patrol the room. Avoid turning the exam into a power trip situation. Just maintain control.

Another aspect of cheating is plagiarism. Plagiarism is the appropriation of another person's words or ideas. It is too large to treat in any detail here. One advantage from your point of view is that you do not have to handle plagiarism in real time. You have the plagiarist's work, together with the putative source material, in front of you. You may consider it carefully, show it to colleagues, ask your undergraduate director how to proceed. The best policy is not to attempt to act alone.

One could easily write another booklet about techniques to catch cheaters. In some departments, exams are xeroxed (or at least a sample of them is xeroxed) before they are returned to students. This is to dissuade students from altering the graded exams and then coming back and requesting more points. Some departments (such as my own) use elaborate statistical procedures to detect unnatural correlations among students' answers on multiple choice exams. A number of other devices are available.

The point is that it is worth spending a few moments thinking about how you will handle cheaters. There are many pitfalls to be avoided. There is nothing very pretty about a potential cheating situation. Just remember that the students have rights that you must respect. Become acquainted with your university's procedures.

4. Frustration

One of the most commonly heard complaints by college mathematics instructors, especially experienced instructors, is this: "Math 297 is a prerequisite for the course that I am teaching yet the students don't seem to know anything from Math 297." A variant of this is "My calculus students cannot add fractions" or "My calculus students don't know how to simplify $(a+b)^2$."

Indeed, these are valid complaints. It is also valid to complain about the high cost of living, or about death and taxes. But these are facts of life and we as math instructors must deal with them. The truth is that instructors think about math all day long, every day. We see the entire curriculum as a piece; for the experienced math instructor, there are no seams and creases between linear algebra, calculus, differential equations, and so forth. We swim effortlessly through the ideas, using whatever tools are needed. [By the way, if this doesn't exactly describe you then don't panic; I'm using a bit of poetic license here.] Students are different. They think about math when they are in the math classroom and (one hopes) for a few designated hours outside the classroom, but they are not inured in the subject.

So what is the point? It is simple. If you are in the second week of freshman calculus and you need to add two algebraic expressions that are fractions, then gently remind the students how to do it. If you need to simplify the expression $(a+b)^2$, then say "You remember how this works—right?" After a few gentle reminders, most of the students fall into the flow and they *will* remember how it goes.

Take a break and watch the "Tonight" show or "Late Night". Listen to the monologue. If the host is going to crack a joke about someone slightly less famous than Bill Clinton, then he gently reminds the audience who it is that he is talking about. It's just good sense. These television hosts can be even less sure of how well informed their audiences are than we can be in our math classes. They guarantee that their viewers will *understand* by providing a bridge.

Contrast these recommendations with the following rather common alternative: the professor needs to add two fractional algebraic expressions, so he/she just barrels through it (rapidly and without comment, as one would do it for a colleague). After a few moments some hands are raised, some hesitant questions are asked, and it soon becomes clear that many students are lost. The professor says "What is the matter with you people? This is high school stuff. Am I a babysitter or what?" (I'm not making this up; I have colleagues who do just this.)

In my view the behavior of the mathematics instructor in the last paragraph is the mathematical analogue of shooting one's self in the foot. The professor (perhaps unconsciously) *sets up* a situation for failure. And there is no good or useful point to it. It not only sets a bad tone for that day, but also for the remainder of the course. It really requires no extra effort to anticipate these little pitfalls, and to devote a few seconds to allaying them. And it does a world of good.

As indicated at the beginning of this section, these frustrations also present themselves at more advanced levels—even with math majors. As an instance, linear algebra is often a prerequisite for multi-variable calculus. And well it should be, for matrix language is a natural vehicle for expressing the derivative, the chain rule, and so forth. But it is an artifact of the American mathematics curriculum that linear algebra is often taught in a vacuum. The students have no hooks to hang the ideas on, and they do not remember them very well. There is no alternative, if you want to keep your multi-variable calculus course on an even keel, to giving a whirlwind review of the salient linear algebra ideas as you use them. Here, by "whirlwind", I mean a five or ten minute snapshot, on the fly, of the relevant idea right before it is used.

The problem described here is one of the few in this booklet that plague the experienced instructor somewhat more than the novice. We tend to lose our patience, and to forget the struggles of the uninitiated. An instructor who has been dealing with, and teaching, the ideas for twenty years cannot understand why students don't remember what they have already seen once. *Once*! The key to success here is to try to develop (or remember) a little sensitivity to the point of view of the students.

5. Difficult Questions

At several junctures in this booklet I have mentioned some spine-tingling, bone-chilling, conversation-stopping questions that students can and will ask. One of these is "What is all this stuff good for?" Another is "Will this be on the test?" Another is "Why don't you prepare your lectures better? You are wasting our time."

If you are asked the third question then the fault probably lies with you. You should have done a better job preparing your lecture. If things are really going dreadfully, you might say to the class "I apologize. This lecture is going very badly. Let's quit for today." Nobody will take this amiss, and it is probably the most diplomatic way out of an uncomfortable situation—but do not use this device more often than about once per semester. The best policy is to use forethought to prevent such an encounter.

Dealing with the first two questions, and others like them, is something that you will learn to do through bitter experience. In America in the 1990s, we endeavor to educate a broad cross-section of the population. We cannot assume, as perhaps a don at Oxford could one hundred years ago, that our students are at the university primarily to learn to become refined citizens—and that they are happy to consider whatever we set before them. In particular, today's students are prone to *challenge* what we are doing. It is a part of your job to be prepared to answer their challenges. The challenges are not generally hostile. But having respect for your audience requires that you be prepared to provide a thoughtful response.

Let us consider the question "Will this be on the test?" If you are going to present something to your students and have no intention of testing them on it, then you have several choices. You can tell them up front that they will not be tested on it; they should just sit back and listen. Or you can tell them that they *will* be tested on it and then, indeed, test them on it. Or you can tell them that they will be tested on it and then *don't* test them on it. Know consciously which choice you have made before you proceed.

If you choose the first route indicated in the last paragraph, you might put the exercise in context like this: Explain that some ideas are difficult and deep. It requires several exposures to them before they begin to make sense. This is an opportunity for the student to begin to ponder something important. Students are pleased to be treated like fellow scholars, and will act accordingly.

The question "What is all this stuff good for" is treated in Section 1.16. You, the mathematician, can get so wrapped up in your mathematics that such a question can catch you entirely off guard. Spend a few moments arming yourself against it.

The main point is this. You are not lecturing fellow mathematicians, who are inured to your point of view. You are lecturing students. Students will challenge you and ask questions. Some of these questions are difficult. If you want to retain the students' respect then you must be prepared to deal with their queries and to understand their point of view.

6. Discipline

One hopes that, in a college environment, discipline will not be a big problem. But there are difficulties that can arise.

In a big class, with a hundred or more people, student talking can get out of hand. Many students read the newspaper, or knit, or eat their lunches, or write letters to their friends. Some students, when the lights are turned low, engage in romantic activities. Students come in late and leave early. Students sleep.

I see no point in making a spectacle over a student who is not causing a disturbance. If a student is quietly eating lunch, that is no problem for me. If a student comes in late or needs to leave early and does so in an orderly and non-disruptive fashion, then I let that student alone. To make a scene will only alienate the whole class.

How should you handle a student who causes a disruption? First try something gentle like "OK, let's quiet down." Another technique is to simply stop talking until you have everyone's attention. If one or two applications of "nice" is ineffectual, then come down on the offender quickly and sternly. Example: "Mr. Herkimer, if you want to talk then please leave the room." or "Ms. Huguenot, everyone else is here to learn. Please keep quiet." If you deliver these injunctions firmly and with confidence, then they will have a chilling effect.

Often you can arrange for the other (non-offending) students to be the bad guys. If you simply sit down and wait for silence and cooperation, then the other students will "shush" the offenders. You can simply provide a modicum of stern looks.

On those absolutely rare occasions when a class is beyond control, you might throw down your chalk and say something like "This class is hopeless. I'm through for today. We'll try again on Friday." I have never used this last device, and I fervently hope that I shall never have to resort to it. But somehow it gives me strength to know that it is at my disposal. If you do take this extreme measure, you had better let your chairman know what you have done.

I have seen large mathematics classes (of about 400 or more students) which looked like a cross between a rock concert and a Heironymous Bosch painting. Private conversations and mini-dramas were taking place all over the room while groups of students roamed the aisles. The professor stood at the front of the room, bellowing away on his microphone, while a small percentage of the students attempted to learn something. Such a situation is plainly unacceptable. Nobody can ask a question, nor can there be any interchange of ideas, in such an atmosphere. But, generally speaking, a situation like this comes about *gradually* over the course of the semester; it happens because someone (most likely the professor) lets it happen.

Some professors prefer to deal directly, and in advance, with the discipline problems connected with large lectures. On the first day of class in a large lecture, these instructors tell the class that large classes present special organizational problems. In order to make the experience as beneficial as possible for everyone, the instructor goes on to prescribe certain rules of behavior in the classroom. These include no eating, no talking, no reading of the newspaper, no coming in late, no leaving early, and so forth. Remember that the first few lectures of your class are your chance to set the tone. You may wish to take the opportunity to "lay down the law" (gently).

Of course if you are going to use the system described in the last paragraph—and it is a perfectly reasonable one—then you must follow through on it. If a student breaks one of the rules you have laid down, then you must call him/her on it: "Mr. Schlobodkin, I said no eating in class." or "Ms. Mergetroyd, if you need to leave class early then you shouldn't come at all."

Because you are an authority in your field, because you give out the grades, and because you hold sway over students' lives, you have both moral and *de facto* authority in the classroom. As a result, if you comport yourself like a concerned, dedicated professional, you should have relatively few disciplinary problems. If you nip disciplinary problems in the bud, and handle them with dispatch, then they will not get out of hand and your classes will go smoothly. But your antennae should be out for trouble. You will figure out quickly who the wise guys and troublemakers are. When they start to rattle, you start to roll.

Always remember that you have the power to command respect, but you cannot *demand it.* If you present the image of an organized, knowledgeable professional who is trying to do a good job of teaching, then most students will play ball with you. If instead you are a bumbling, unprepared clod who clearly doesn't care a damn for the class, then you can expect like treatment from the students.

I used to have a colleague who handled late arrivals to a large lecture in the following fashion. He would cease to lecture and make a show of timing how long it took the student to get seated. Then he would say: "There are 150 students in this class. It took you ninety seconds to get seated. Thus you wasted 3.75 hours of their time. Now each student here has paid x number of dollars to take this course. Let us next calculate how many of their dollars you have wasted." And so forth.

Let us consider the effect of this practice. Certainly any student who planned to come in late in the future would wear a disguise. But I cannot help but think that this sort of arrogant behavior on the part of the instructor suggests a serious attitude problem. Students will lose respect for an instructor who behaves in this fashion.

As the instructor in a classroom, you are in charge. You have every right to demand a certain type of behavior from students, and to enforce discipline. But you must not, in the course of disciplining a student, diminish his/her self-respect. Students are young adults, and should be treated as such.

We have all seen parents who cannot control their children. We have also seen 95 pound fathers who hold tremendous sway over their 250 pound linebacker sons. There is a difference between *demanding* respect and *commanding* respect. The first is easy, and often doesn't work. The second is an art that you need to cultivate. The techniques suggested in this booklet should help you in this task.

Remember this: you should be conscious of maintaining discipline in a class from day one. This is not to say that you should be an unbending authoritarian; far from it. But if you let a class slide out of control for six weeks and then try to use the techniques suggested here (or other techniques) to take back the reins of power then you will have an extremely difficult and unpleasant time.

The famous mathematics teacher R. L. Moore is said to have once brought a Colt 45 to an unruly math class, set it conspicuously on the table, and then proceeded into his lecture in a room so quiet that one could have heard hair grow. This technique may have been suitable in Texas fifty years ago. These days, however, I would recommend the use of more civilized techniques for keeping order.

7. Mistakes in the Lecture

The most important rule to follow before giving a lecture is to prepare. How much you prepare will depend on you: on your experience, your confidence, your training, and so forth. Being fully prepared gives you the flexibility to deal creatively with the unexpected.

But nobody is perfect. No matter how well prepared you are, or how careful, you will occasionally slip up. In the middle of a calculation, a plus sign can become a minus sign. An x may become a y. You will say one thing, think a second, write a third, and mean a fourth. It is best if you can handle these slips with a flair, and particularly without sending the class into a tailspin.

I endeavor in my classes to create an atmosphere in which students are comfortable to shout out "Hey, Krantz, you forgot a minus sign." Or "Is that a capital 'F' or a lower case 'f'?" This is a form of participation, and a very constructive one. If you handle these situations badly, then students will be less inclined to ask questions or to approach you on other, more important matters.

If mistakes are small, and occur in isolation, then they will not damage the learning process. But if they are frequent or, worse, if they snowball, then you will lose almost everyone, give a strong impression of carelessness, set a bad example, and (to oversimplify) turn off the class.

You may endeavor to bail out of an example that you are lousing up by saying "Well, this isn't working out. Let's start another example." It won't work. This is in the vein of two 'wrongs' not making a 'right'. The only solution here is not to make mistakes and to handle those that you make anyway with a flair.

However: If you can see that the example you are working on is getting out of control, if you *know* that it is going from bad to worse, what do you do? Do not spend the rest of the hour trying to slug it out. This is uncomfortable, counterproductive, and will not teach anyone anything. Instead apologize, say that you will write up the solution and hand it out next time, and move on. This advice may seem to fly in the face of Section 2.8, and to contradict the last paragraph, but it is only meant for extreme situations. Making mistakes is one of the surest ways to lose control of a class. It is the mathematical analogue of an equestrian letting go of the reins. Strive not to do it.

Besides preparing well, there are technical devices for minimizing the number of errors that you make. When I am working an example in a lower division class, I pause *frequently* to say "let's make sure this is right" or "let's double check this step." I often pick out a student (who I know will respond well) and ask him/her whether that last step was done correctly. This procedure provides a good paradigm for the students; it also allows note takers to catch up and allows the bright students to strut their stuff in a harmless manner.

One of the most common ways that students make mistakes in their work is by trying to do too much in their heads. Set a good example. Write out all calculations. Point out *explicitly* that you have had many years of experience with this material yet you still use lots of parentheses and write out every step.

8. Advice and Consent

If your students take a shine to you, and many of them will, then they will view you as looming larger than just "the math teacher." They will come to you for advice on all sorts of things, from the purchase of a computer or symbol manipulation software, to advice on the purchase of a car, to advice on how to handle their parents, or advice on very private matters.

A good rule of thumb for you as teacher is to stick to things that you know. You are probably well qualified to give guidance about math books, which section of calculus to sign up for next semester, which computer to buy for which purposes, or whether MACSYMA is preferable to MATHEMATICA. If you are an auto buff you could give advice about wheels. But being in a position of authority and being asked for advice by a semi-worshipful student is heady stuff, and you had better be careful.

When you are advising students as to which math class to take, it is easy to fall into the trap of unintentionally (or, more is the pity, intentionally) criticizing your colleagues. The practice of this indiscretion is unfortunately rather common. Please do not fall into it.

The students are your clients, but in some sense you work for their parents (since they probably pay the freight). You are almost certainly out of place to advise your students on how to behave as children. Do so at your own risk.

When a student starts asking you about private matters then you are in dangerous territory. It is often difficult to discern the difference between (i) a student asking how to deal with a significant other and (ii) a student making a come on. I hate to keep mentioning sexual harassment and political correctness, but they are a part of life these days. Defending yourself against an allegation of either is one of the loneliest and most miserable battles that you might ever have to fight. It can threaten your self-respect, your career, and your marriage. A word to the wise should be sufficient in this matter. Section 3.9 treats sexual harassment and related topics in more detail.

Many undergraduates enjoy having a faculty member as a friend. If you are open to it, you can have two or three students hanging about your office at just about any time of the day (or night). It's a boost to the ego to have these attentions, young people are often quite refreshing, and this device provides a convenient way to rationalize wasting a heck of a lot of time. You will have to decide for yourself how you want to handle this trap.

Consider this a bit differently: If you make yourself available all day long to help your students with their math (never mind getting involved in their personal lives), you will indeed attract customers. Make yourself available to other professors' students and you will have even more customers. You can provide tutorials, make up extra homework assignments for students, and find innumerable other ways to while away the day. But the set of activities that contribute to a successful academic career and the set of activities that I have just described have a rather small intersection. That sounds rather Machiavellian, so let me be more gentle: Almost all learning is ultimately accomplished by the individual. I've engaged in almost all the activities described here; they have produced few lasting results and, in the end, have truly helped very few students.

Most of those who end up surviving in the academic game are people who decide that, no matter how much they love their students, they love themselves a bit more. Your students won't like you any the less for saying "I have to do some work now; let's talk at another time." You will figure this out for yourself eventually, but you heard it here first.

Finally, all of the advice in the last three paragraphs must be filtered through the value system of the institution at which you teach. Swarthmore takes a different approach to education than does M.I.T.—and both are excellent. Do be sensitive to what is expected of faculty at your institution.

9. Sexism, Racism, Misogyny, and Related Problems

Nobody, certainly no educated person, thinks of himself/herself as a sexist, or a racist, or a misogynist. That is what is so insidious about these sins when they come up in a university environment.

I don't want to preach about any of these topics; rather I would like to mention some pitfalls. Common complaints from students are these: (1) the professor calls on male students more than on female student, (2) the professor will not answer questions from female students, (3) the professor shows favoritism towards female students in his/her grading policies, (4) the professor talks down to minority students, (5) the professor seems to believe that women are less able than men, (6) the professor seems to believe that Caucasians are more able than minorities, (7) the professor demands more from Asian students, (8) the professor makes suggestive cracks in class.

None of these complaints is cooked up. They have all been tendered, very seriously, by genuinely outraged students. It goes without saying that in all instances "male" may be switched with "female", "Caucasian" may be switched with "minority", and so on. The issue here, and I cannot emphasize this too strongly, is not that any particular class of people is persecuting any other particular class of people. It is rather that every instructor has his/her foibles and shortcomings and biases and these will often be perceived by students through the filter of whatever issues are currently in the air.

Of course it is sometimes the case that a student with one of the complaints described above is doing poorly in the course and is looking for an excuse or a scapegoat. It is also the case that sometimes the professor just doesn't realize that he/she is behaving in a manner that some students take amiss. In short, there is plenty of room for misunderstanding.

It is pretty obvious that you shouldn't touch your students. An arm around the shoulder or even a prolonged and enthusiastic handshake can be easily misinterpreted. But many people, especially those new in the teaching profession, are not aware of the subtleties involved in the legal definitions of sexism, racism, and so forth. Let me surprise you with some other aspects of harassment that you may not know:

If you are in the habit of saying "I don't want to blow smoke up your dress" or "this is a pregnant idea" or "this problem is a bitch" or "bulls—" during your lectures then, by a strict interpretation of the statute, you may be guilty of sexual harassment. This is true regardless of your sex or the sex of the members of your audience. In fact if you even tolerate this language from others who are in the classroom, then you also may be guilty! The spirit of the law is that if someone *feels* offended then they are offended.

If you have in your office a Playboy pin-up, or a (closed, unread) copy of *Playboy* sitting around where others can see it, or a *Frederick's of Hollywood* catalog, or a Chippendale's poster, then you may be guilty of sexual harassment.

You may feel that being forced to monitor your language or other behavior this closely is an abridgment of your First Amendment rights, and you may be correct in this feeling. My view, much as I hate censorship, is that there is no percentage in going through life wearing a 'kick me' sign. You can use this language, or display these artifacts, all you like when alone or in private circumstances. As a teacher you are something of a public figure and must suffer certain restrictions.

A number of universities in this country have distributed detailed guidelines to their faculties about the issues being discussed here. Some have gone further, and indicated specific words or phrases that ought not to be used. For instance, you should not say "snowman" but should use a suitably laundered asexual alternative. Various standard English phrases are suggested to be off limits and substitutes are recommended. The point is that we are dealing with very delicate and emotion-charged issues of legality and morality here and one has no choice but to take them seriously.

It is no fun having to deal with issues of racism or sexism or misogyny. A complaint lodged against you is, in effect, an attack on your character and your integrity. So be aware that these areas are a potential problem for all of us.

Behave accordingly. If you are called on the carpet for any of these matters, do not become defensive. Show respect for the complainant. Get help from your department head or from your dean of human resources. This is serious business.

10. Begging and Pleading

Some students will come to you with unreasonable requests. They will tell you, after doing poorly on an exam, that if they do not pass this test, or this course, with a certain grade then they cannot continue in the pre-med program, or the microbiology program, or whatever. Of course you, as professor, can verify rather quickly whether this claim is true. But it does not matter. If the test was so important to the student then the student should have studied harder. The student should have come to your office hours for help before the test. If the student's homework was weak then the student should have seen the writing on the wall. Be cautious of these pleas. While you do not want to be heartless and unsympathetic, you also do not want to find yourself gradually being drawn into an ever more complicated morass of tricky moral dilemmas.

A desperate student will offer you all sorts of inducements to change grades. Discretion prevents me from enumerating what some of these may be, but they range from the pecuniary to the personal. *You must brush these attempted bribes off with the disdain that they deserve.* If you act as though you are considering and then rejecting them, then you are looking for trouble.

It really is true that if you look and/or act like a student then students will find you more approachable. They will come to you more readily with propositions that they wouldn't consider broaching with a more detached faculty member. In short, younger faculty are more vulnerable. This is one reason for dressing differently from students and maintaining a slight distance. Again, this may sound cold. But I speak here from hard personal experience.

As has been mentioned elsewhere, you must be sensitive to sexual harassment issues. This is not a pretty subject, but acting receptive—even mildly so—to any proffered inducements is only looking for trouble.

Some students will ask to be given the opportunity to take a test again. You simply cannot allow this. For one thing, it is unfair to the other students. Second, if the others find out then they will become angry—and justifiably so. You *can* give the student a second try *ex officio* and go over the test afterwards with the student. This can be a device for giving the student some encouragement; you might tell the student that "I can see from this unofficial exam that you know the material better than your official exam suggests. If you do well on the final then you can still probably get a grade of 'B'." However you must engage in this charity sparingly, if for no other reason than it can use up large chunks of your time. Also, it is too easily misinterpreted.

A favorite student response to a poor test grade is "I did very well on the homework but my test grade does not reflect what I know." Of course some students may "clutch," or panic on a test. As a student, I have done so myself. But unfortunately many students do their homework by copying examples from the text, or the lecture—merely changing the appropriate numbers. This leads to a minimum of understanding. You must stress to your students that, when they study for a test, they should reach proficiency *without* recourse to the book or notes. If the book or notes are necessary, then the technique has not been mastered.

One of the most common student remarks is "I really understand the material but I cannot do the problems." A variant is "I can do the problems on the homework but I cannot do the problems on the test." Consider if you will these analogous statements: "I really understand how to swim but every time I get in the water I drown." and "Playing the piano sure looks easy when Arthur Rubinstein does it. I wonder why I cannot do it."

If you are a good lecturer, then you will make the material look easy, or at least straightforward. This can lull students into a false sense of security. You must continually warn them of the importance of mastering the material themselves—and of *practicing.* And this point leads to the second ludicrous statement recorded above. The fact that a student can do the homework problems is meaningless *unless* the student can do them cold, and with the book closed.

This last point seems so obvious that it hardly bears mention. But recall that this is a book about the obvious, and this point bears not only mention but repeated mention. Many students view the learning process as a passive one. You must constantly remind them that this attitude will not do. You can remind them by just telling them. Or you can remind them by giving pop quizzes. Or you can remind them by giving an exam and watching them flunk. But, one way or the other, you must attempt to break through this psychological impasse.

In a related vein, many students think that (i) studying and (ii) just sitting in front of the book are one and the same thing. We all know that studying requires discipline, tenacity, and hard work. It is not the same as getting a massage. Helping your students to understand this point begins with being conscious of the problem.

In my opinion most students want to be told what to do. Your job as teacher is to tell them. Don't make them guess what are the important topics in your class. Tell them. Don't make them guess what they will be tested on. Tell them. Don't make them guess how to study for an exam. Tell them. Don't make them guess what are the pitfalls in studying. Tell them. Is there any reason not to do this? Would you rather deal with the begging and the pleading?

11. Closing Thoughts

Sometimes the easiest way out, when we are faced with some difficult or distasteful task to perform, is to resort to cowardice. We are all guilty of this. At one time or another we have all lied or engaged in subterfuge to avoid unpleasantries.

I do not wish to dwell here on human frailties. But I think that the method that many of us choose to teach—and I have been guilty of this to a degree with certain classes that I really did not want to be teaching—is a form of cowardice. We just skulk into the room, write the words on the board, and convey with body language and voice and attitude that we are not interested in questions or in much of anything else connected with this class. Then we turn tail and skulk out of the room. I was once told (tongue-in-cheek, I think) that the secret to success in undergraduate teaching is "never let a student get between you and the door". Not an admirable attitude, but one that many of us have held from time to time.

This booklet has been an effort to fight this form of cowardice, both in myself and in others. Teaching can be rewarding, useful, and fun. To make it so does not require an enormous investment of time or effort. But it does require that you have a proper attitude and that you be conscious of the pitfalls. It does require being sufficiently well prepared in lecture so that you can concentrate on the *act* of teaching rather than on the epsilons. And it requires a commitment.

We must believe that being a good teacher is something worth achieving. We must provide some peer support to each other to bring about this necessary positive attitude towards teaching. The last thing I want is for mathematicians to spend all day in the coffee room debating the latest pedagogical techniques being promulgated by some Ivy League school of education. I want to see mathematicians learning and creating mathematics and sharing it with others. But those others should include undergraduates. That is what teaching is about.

Bibliography

[AMH] J. Callahan and K. Hoffman, et al, Amherst project calculus manuscript, 1991.

[CTUM] Committee on the Teaching of Undergraduate Mathematics, *College Mathematics: Suggestions on How to Teach it*, Mathematics Association of America, Washington, D.C., 1979.

[HAL] D. Hughes-Hallet, et al, *Calculus*, John Wiley and Sons, New York, 1992.

[HOF] D. Hoffman, The Computer-Aided Discovery of New Embedded Minimal Surfaces, *Math. Intelligencer* 9(1987), 8-21.

[KIR] W. Kirwan, et al, *Moving Beyond Myths*, National Research Council, The National Academy of Sciences, Washington, D.C., 1991.

[KUM] Kumon Educational Institute of Chicago, assorted advertising materials, 112 Arlington Heights, Illinois, 1992.

[REZ] B. Reznick, *Chalking it Up*, Random House/Birkhauser, Boston, 1988.

[ROS] A. Rosenberg, et al, *Suggestions on the Teaching of College Mathematics*, Report of the Committee on the Undergraduate Program in Mathematics, Mathematics Association of America, Washington, D.C., 1972.

[RUD] W. Rudin, *Principles of Mathematical Analysis*, 3rd. Ed., McGraw-Hill Publishing, New York, 1976.

[STE] L. Steen, et al, *Everybody Counts*, National Research Council, The National Academy of Sciences, Washington, D. C., 1989.

[SYK] C. Sykes, *Profscam*, St. Martin's Press, New York, 1988.

[THU] W. Thurston, Mathematical Education, *Notices of the A. M. S.* 37(1990), 844-850.

[TRE] U. Treisman, Studying students studying calculus: a look at the lives of minority mathematics students in college, *The College Mathematics Journal*, 1992, 362-372.

Index